AGRICULTURAL RESEARCH UPDATES

VOLUME 5

AGRICULTURAL RESEARCH UPDATES

Additional books in this series can be found on Nova's website
under the Series tab.

Additional e-books in this series can be found on Nova's website
under the e-book tab.

AGRICULTURAL RESEARCH UPDATES

VOLUME 5

PRATHAMESH GORAWALA

AND

SRUSHTI MANDHATRI

EDITORS

nova
publishers

New York

NOTICE TO THE READER

Library of Congress Cataloging-in-Publication Data

ISSN: 2160-1739

ISBN: 978-1-62618-723-8

Published by Nova Science Publishers, Inc. † *New York*

CONTENTS

Preface   **vii**

Chapter 1   Competitiveness of Bulgarian Farms in Conditions
                  of EU CAP Implementation   **1**
                  Hrabrin Bachev

Chapter 2   Perennial Weeds in Argentinean Crop Systems:
                  Biological and Ecological Characteristics
                  and Basis for a Rational Weed Management   **43**
                  Marcos E. Yanniccari and Horacio A. Acciaresi

Chapter 3   Anti-inflammatory Effect of the Waste Components
                  from Soybean (*Glycine Max* L.) Oil Based
                  on DNA Polymerase λ Inhibition   **63**
                  Yoshiyuki Mizushina , Yoshihiro Takahashi,
                  Isoko Kuriyama and Hiromi Yoshida

Chapter 4   Dry Matter Production, Yield Dynamics and Chemical
                  Composition of Perennial Grass and Forage Legume
                  Mixtures at Various Seeding Rate Proportions   **85**
                  Tessema Zewdu Kelkay

Chapter 5   Genetic Variation of Globulin Composition in Soybean Seeds   **101**
                  Kyuya Harada, Masaki Hayashi and Yasutaka Tsubokura

Chapter 6   Different Effects of Garlic Preparations: A Mini Review   **117**
                  Luziane P. Bellé and Maria Beatriz Moretto

Chapter 7   Vegetable Soybean (Edamame): Production, Processing,
                  Consumption and Health Benefits   **129**
                  Bo Zhang, Cuiming Zheng, Ailan Zeng
                  and Pengyin Chen

Chapter 8   Non-Significant Interactions between Treatments
                  and a Suggested Statistical Approach for Dealing
                  with These Statuses   **139**
                  Zakaria M. Sawan

Chapter 9   Rapid Methods for Improving Nutritional Quality
                 of Hydroponic Leafy Vegetables Before Harvest   **147**
                 Wenke Liu, Lianfeng Du and Qichang Yang

Index   **155**

PREFACE

This compilation examines agricultural research from across the globe and covers a broad spectrum of related topics. In this book, the authors discuss research including the competitiveness of Bulgarian farms in conditions of EU CAP implementation; perennial weeds and management in Argentinean crop systems; the anti-inflammatory effects of the waste components from soybeans; dry matter production, yield dynamics and chemical composition of perennial grass and forage legume mixtures at various seeding rate proportions; genetic variations of globulin composition in soybean seeds; different effects of garlic preparations; vegetable soybean (edamame) production; the effect of nitrogen (N) fertilizer and foliar application of potassium (K) and mepiquat chloride (MC) on the yield of cotton; and rapid methods for improving the nutritional quality of hydroponic leafy vegetables before harvest.

Chapter 1 - This chapter suggests a holistic framework for assessing farm competitiveness, and analyses competitiveness of different type of Bulgarian farms during EU CAP implementation. First, it presents a new approach for assessing farm competitiveness defining farm competitiveness and its three criteria (efficiency, adaptability and sustainability), and identifying indicators for assessing the individual aspects and the overall competitiveness of farms. Next, it analyzes evolution and efficiency of farming organizations during post-communist transition and EU integration in Bulgaria, and assesses levels and factors of farms competitivenessin the conditions of CAP implementation. Third, it assesses the impact of EU CAPon income, efficiency, sustainability, and competitivenessof Bulgarian farms.

Chapter 2 – In the last two decades, cropland changes from conventional tilled systems to zero tillage systems have been detected in Pampean agroecosystems. Dominant species are the primary weed due to the fact that they are adapted to the cropping system. Weed population shifts were observed when conventional tillage systems were changed to non-tillage. However, a great proportion of perennial species would be expected in non-till environments. In Argentinean crop systems only a few perennials have transcended in importance.

In this context, *Sorghum halepense* has been one of the most important species in dispersion and aggressiveness. In addition, the low sensitivity of several populations to glyphosate contributes to their complex management. Populations from diverse ecological regions have differential mechanisms for adaptation according to the different environments where they have been growing.

Even in non-till systems, *Cynodon dactylon* is another summer crops primary perennial weed and it is considered as highly plastic species in response to growth factors. Water and radiation have been major factors affecting the growth, conditioning the aggressiveness of the weed.

Recently, the combining uses of zero tillage and round-up ready soybeans have promoted an increased shift in weed herbicide resistance. The detection of a glyphosate-resistant *Lolium perenne* population has increased interest in researchers of these biotypes. Several physiological traits influence their management in winter cereal crops.

The plasticity of these weeds to adapt to zero tillage systems requires a design of integrated weed management program according to biological traits of the weeds. Several strategies of mapping and monitoring of weed populations, crop rotations, biocontrol and rotation of herbicides, among other, could be used to maintain the competitive ability of crops rather than eradication of weeds.

Chapter 3 – During screening for selective DNA polymerase (pol) inhibitors, the authors purified compounds 1–3 from the waste extracts of soybean (*Glycine max* L.) oil (i.e., the gum fraction). These isolates were identified by spectroscopic analyses as glucosyl compounds, an acylated steryl glycoside [β-sitosteryl (6'-*O*-linoleoyl)-glucoside, compound 1], a steroidal glycoside (eleutheroside A, compound 2), and a cerebroside (glucosyl ceramide, AS-1-4, compound 3). Compound 1 exhibited a marked ability to inhibit the activities *in vitro* of mammalian Y-family pols (pols η, ι, and κ), which are repair-related pols, and observed the 50% inhibitory concentration (IC_{50}) to be 10.2–19.7 μM. Compounds 2 and 3 selectively inhibited the activity of eukaryotic pol λ, which is a repair/recombination-related pol, with IC_{50} values of 9.1 and 12.2 μM, respectively. However, these three compounds did not influence the activities of the other mammalian pol species tested. In addition, these compounds had no effect on prokaryotic pols or other DNA metabolic enzymes, such as human DNA topoisomerases I and II, T7 RNA polymerase, T4 polynucleotide kinase, and bovine deoxyribonuclease I. Compounds 1–3 also exhibited no effects on the proliferation of several cultured human cancer cell lines. Compounds 2 and 3 suppressed the inflammation of mouse ear edema induced by TPA (12-*O*-tetra-decanoylphorbol-13-acetate); therefore, the tendency for pol λ inhibition *in vitro* by these compounds showed a positive correlation with anti-inflammation *in vivo*. These results suggested that these glucosyl compounds from soybean waste extracts might be particularly useful for their anti-inflammatory properties.

Chapter 4 - Dry matter (DM) yield, relative yield, relative total yield, aggresivity index, relative crowding coefficient and chemical compositions of grass-legume mixtures were studied in a field experiment at Haramaya University field station, Ethiopia in 2005 and 2006. *Chloris gayana, Panicum coloratum, Melilotus alba* and *Medicago sativa* were planted as pure stands and in mixtures using 50:50, 67:33, 33:67, 75:25 and 25:75 seed rate proportions of grasses and legumes, respectively. *C. gayana* mixed with *M. alba* at seed rates of 50:50 and 33:67 produced DM yields of 26.7 and 26.1 t ha^{-1}, respectively, which were higher than other mixtures and pure stands. The total yields of all grass-legume mixtures were greater than monocultures. The mean relative crowding coefficient values of the grasses in the mixtures with legume species were high, indicating that they contributes to higher DM yields of the grass-legume combinations. Forage legume monoculture and legume-grass mixtures had higher crude protein contents than pure grass stand, whereas grass monocultures had higher fibre fraction contents compared to legume monocultures and legume-grass mixtures. *C. gayana* mixed with *M. alba* at seeding rate of 50:50 and 33:67 can be introduced to alleviate the feed shortages into smallholder farms in the eastern parts of Ethiopia. Further studies on animal performance should be conducted using feeding experiments and under grass-legume mixed pasture grazing to confirm the performance of these mixtures under farmers conditions.

Chapter 5 - About seventy per cent of total protein in soybean seeds consists of two major storage proteins, glycinin (11S globulin) and β-conglycinin (7S globulin). Glycinin is composed of five subunits, $A_{1a}B_{1b}$, $A_{1b}B_2$, A_2B_{1a}, (group I), A_3B_4 and $A_4A_5B_3$ (group II). Beta-conglycinin is composed of three subunits, α, α' and β.

A mutant line lacking β-conglycinin was obtained by γ-irradiation. The deficiency was controlled by a single recessive allele, *cgdef* (*β-conglycinin-deficient*). The result of Southern and northern blot analyses indicated that the deficiency is not caused by a lack of, or structural defects in the β-conglycinin subunit genes, but rather arise at the mRNA level. It was assumed that the mutant gene encodes a common factor that regulates expression β-conglycinin subunit genes. Another β-conglycinin deficient mutant QT2 was identified from a wild soybean. The phenotype was found to be contolled by a single dominant gene *Scg-1* (*suppressor of β-conglycinin*). The physical map of the *Scg-1* region covered by lambda phage genomic clones revealed that the two α subunit genes were arranged in a tail-to-tail orientation, and the genes were separated by 197 bp in *Scg-1* compared to 3.3 kb in the normal allele. Moreover, small RNA was detected in immature seeds of the mutants by northern blot analysis using an RNA probe of the α subunit. These results strongly suggest that β-conglycinin deficiency in QT2 is controlled by post-transcriptional gene silencing through the inverted repeat of the α subunits.

A glycinin deficient line was developed by combining $A_4A_5B_3$ subunit deficiency, an A_3B_4 deficient line in wild soybean and a group I deficient line produced by γ-irradiation. Each deficiency was controlled by a single recessive gene.

A line lacking glycinin and β-conglycinin was generated by combination of a glycinin deficient line and the *Scg-1* gene. This line grows normally and contains large amount of free amino acids, especially arginine.

In this chapter, the authors will describe the characteristics of these lines and discuss their uses and significance in soybean breeding.

Chapter 6 - Currently, people look for alternative methods for the treatment of diseases, especially those coming from the aging. Garlic (Allium sativum), is a plant known since ancient times that has been used to treat a large number of diseases in popular medicine. There are different types of preparations (raw, extracts, oil and powder), and each one has specific effects according to the predominant components. These compounds stem from the way that the preparation is obtained. Therefore, this study was aimed at discussing the beneficial and side effects of the use of garlic preparations and garlic bulbs in the alimentation.

Chapter 7 - Soybean is primarily grown as a commodity crop for its high protein and oil content in seed. In recent years, tremendous interests have been generated among consumers in soyfoods including tofu, natto, soymilk, soy-sprouts, soy-nuts, and edamame due to their nutritional value and health benefits. Edamame is a vegetable soybean harvested at full green pod stage, processed as in-pod or shelled seed product, marketed as fresh or frozen produce, and consumed in various forms of stews, salads, dips or salted snacks. As a rich source of protein, vitamins, calcium and isoflavones, edamame is a favorable food product for prevention of cardiovascular disease, mammary and prostate cancers, and osteoporosis. As a conventional (non-GMO) or organic crop, edamame requires specific varieties with desired seed size and quality attributes, specific cultural management practices for optimal production, and special processing procedures for quality product development. As consumers around the globe, particularly in the west, continue to incorporate edamame into their diets as healthy food products, edamame will rapidly expand in the market place. For the producers, edamame as a cash crop has significant potential for substantial profits because of the high market value. This chapter provides an overview on edamame production including current production status and factors affecting production, processing procedures and conditions, consuming exploration and consumer acceptance, as well as the health benefits with an emphasis on nutrition.

Soybean [*Glycine max* (L.) Merr] was introduced into the US in the early 1800's as a forage crop (Mease, 1804). Subsequently, it was transformed into a major commodity crop for oil and meal production. Soybean has also served as a relatively minor food crop for a long time. However, food-type soybean industry has spent 20 years growing from door to door sale to abundant supply throughout the US (Soyfoods Association of North America, 2011). The consumption of soyfood has increased not only because soybean is an important source of complete protein, but also because soyfood helps humans to reduce cardiovascular disease, osteoporosis, and cancer risks (Messina, 1999; Messina, 2009). Soyfood is typically classified into two categories based on seed size: small- and large-seeded. Tofu (soybean curd), edamame (vegetable soybeans), miso (fermented soup-base paste) and *soymilk* (soybeans soaked, ground fine, and strained) are made from large seed (> 20g/100 seeds), but for natto (fermented whole soybeans), soy sauce (tamari, shoyu, and teriyaki), and soy sprouts, small seeds (< 12g/100 seeds) are desirable (Zhang et al., 2011).

Vegetable soybean, known as *edamame* in Japan and *maodou* in China (Shurtleff and Lumpkin, 2001), is harvested at R6 to R7 growth stage when seed are still green (Fehr et al., 1971). Edamame is usually eaten as a snack in Japan and China after being boiled in salt water. It can also be stir-fried or boiled to add into stew or soup similar to sweet pea or lima bean (Mebrahtu and Devine, 2009). In addition to large seed size, edamame is desired to have high sugar content, smooth texture, nutty flavor, and lack of beany taste (Young et al., 2000; Mohamed and Rao, 2004; Mebrahtu and Devine, 2009). Recently, research on edamame has been focusing on breeding development including seed quality traits and diversity analysis, production improvement, consuming exploration, and marketing evaluation.

Chapter 8 - A field experiment was conducted to study the effect of nitrogen (N) fertilizer and foliar application of potassium (K) and Mepiquat Chloride (MC) on yield of cotton. Seed cotton yield per plant and seed cotton and lint yield per hectare; have been increased due to the higher N rate and use of foliar application of K and MC. No significant interactions were found among the variables in the present study (N, K and MC) with respect to characters under investigation. Generally, interactions indicated that, the favorable effects ascribed to the application of N; spraying cotton plants with K combined with MC on cotton productivity, were more obvious by applying N at 143 kg per hectare, and combined with spraying cotton plants with K at 957 g per hectare and also with MC at 48 + 24 g active ingredient per hectare. Sensible increases were found in seed cotton yield per hectare (about 40%) as a result of applying the same combination.

However, this interaction did not reach the level of significance, so, statistical approach for dealing with the non-significant interactions between treatments, depending on the Least Significant Difference values has been suggested, to provide an opportunity to disclosure of the interaction effects regardless of their insignificance. As a matter of fact the original formula used in calculating the significance of interactions suffers a possible shortage, which can be eliminated through applying the new suggested formula (Sawan, 2011).

Chapter 9 - Hydroponics is an important cultivation method to produce leafy vegetables in protected facilities with many of advantages compared with soil culture. However, some nutritional quality problems often occur for hydroponic leafy vegetables cultivated under cover due to heavy nitrogen fertilizer supply and low light intensity, e.g. high level of nitrate accumulation, low content of vitamin C and soluble sugar and so on. In this chapter, two established short-term methods, i.e. nitrogen deprivation method and short-term continuous lighting method, to improve nutritional quality of hydroponic leafy vegetables before harvest were summarized. Additionally, the regulation strategy for promoting nutritional quality of hydroponic leafy vegetables before harvest and its advantages were discussed.

In: Agricultural Research Updates. Volume 5
Editors: P. Gorawala and S. Mandhatri

Chapter 1

COMPETITIVENESS OF BULGARIAN FARMS IN CONDITIONS OF EU CAP IMPLEMENTATION

*Hrabrin Bachev**

Institute of Agricultural Economics, Sofia, Bulgaria

ABSTRACT

This chapter suggests a holistic framework for assessing farm competitiveness, and analyses competitiveness of different type of Bulgarian farms during EU CAP implementation. First, it presents a new approach for assessing farm competitiveness defining farm competitiveness and its three criteria (efficiency, adaptability and sustainability), and identifying indicators for assessing the individual aspects and the overall competitiveness of farms. Next, it analyzes evolution and efficiency of farming organizations during post-communist transition and EU integration in Bulgaria, and assesses levels and factors of farms competitiveness in the conditions of CAP implementation. Third, it assesses the impact of EU CAP on income, efficiency, sustainability, and competitiveness of Bulgarian farms.

Keywords: Efficiency, adaptability, sustainability, and competitiveness of farms, transitional agriculture, EU integration, impacts of EU CAP, Bulgaria

1. INTRODUCTION

The issue of farm competitiveness is among the most topical in academic, business and political respect. There have been numerous studies on competitiveness of different type and kind of farms in developed, transitional and developing countries [Benson; Delgado et al.; Farmer; Fertő and Hubbard; Mahmood; Popovic et al.; Pouliquen; Shoemaker et al.; Zawalinska]. Nevertheless, up to date, there is no widely accepted and comprehensive framework for assessing farm competitiveness in different market, economic, institutional and

---

* Correspondence address:, e-mail: hbachev@yahoo.com.

natural environment. Usually farm competitiveness is not well defined and it is studied through traditional indicators of technical efficiency, productivity, profitability etc. At the same time, important aspects of farm competitiveness such as the governance efficiency, the potential and incentives for adaptation, and the sustainability are commonly ignored in the analyses. Furthermore, with very few exceptions [Bachev 2010b; Koteva and Bachev] there are no comprehensive studies on farm competitiveness in Bulgaria during EU integration and CAP implementation. This chapter suggests a holistic framework for assessing farm competitiveness, and analyses competitiveness of different type of Bulgarian farms during EU CAP implementation. First, it presents a new approach for assessing farm competitiveness defining farm competitiveness and its three criteria (efficiency, adaptability and sustainability), and identifying indicators for assessing the individual aspects and the overall competitiveness of farms. Next, it analyzes evolution and efficiency of farming organizations during post-communist transition and EU integration in Bulgaria, and assesses levels and factors of competitiveness of different type of farms in the conditions of CAP implementation. Third, it assesses the impact of EU CAP on income, efficiency, sustainability, and competitiveness of Bulgarian farms.

2. FRAMEWORK FOR ASSESSING FARM COMPETITIVENESS

2.1. Definition of Farm Competitiveness

Farm competitiveness characterizes the *ability (internal potential, incentives) of a farm to compete on (a) market successfully* [Bachev 2010b]. It is a feature only of the *"market* farms" whatever their specific type is – semi-subsistence (semi-market) holdings, family farms, cooperatives, business enterprises etc. If a farm is non-market (e.g. subsistence holding, member oriented cooperative), or it is quasi or entirely integrated in a larger venture (e.g. processing enterprise, food chain, restaurant, eco-tourism etc.) it has no such attribute.

A good competitiveness means that a farm can produce *and* sell out its products and services *effectively*. The later could be a result of the competitive *prices, variety, quality, time of delivery, location* or other *specificity* (such as newest, uniqueness, organic character, origin etc.) of farm and/or its products. Contrary, the insufficient competitiveness indicates that a farm is experiencing serious problems in producing and marketing its output effectively (or at all) because of the high production *and/or* transaction costs. The farm competitiveness usually refers to farm's ability to compete on a *certain market(s)* – retail, wholesale, local, regional, international, niche, for commodities for direct consumption or processing, mass or specific products, services, etc.

In some cases, a *segment* of farm's activity could be competitive while other(s) not. For instance, in many mix Bulgarian farms the crop production is usually highly competitive while livestock operations are not. Besides, there are various reasons for keeping "profitable" *and* "unprofitable" activities within a farm – e.g. preferences, internal use of "free" resources, technological and transaction costs economies of scale and scope, interdependency of assets or activities, risk management etc. [Bachev 2004, Bachev 2012b]. Therefore, farm efficiency and competitiveness characterize the overall rather than the partial performance of a farm.

The *level* of competitiveness of a particular farm depends on two groups of *factors*:

- *internal factors* - managerial capital, owned resources, potential for innovation and adaptation, productivity, relative power, location, relation specific capital, reputation etc. and
- *external factors* - evolution and maturity of agrarian markets, number and power of competitors, development of downstream and upstream industries, level of public support to agriculture, institutional restrictions, border control measures, liberalization of local markets and international trade etc.

The specific level of competitiveness of a particular farms, or farms in individual sub-sectors, regions and countries depends on internal and outside factors. However, the farm competitiveness is always a *characteristic of the farm* and expresses its *internal potential* (ability) to compete successfully in the *specific* economic, institutional etc. environment. Farm competitiveness is usually assessed in a *relative* term (comparing to other similar farms) or *absolute* term (comparing to other competitors on a market). A particular farm could have a higher, average or lower performance than the other similar farms, and be competitive or uncompetitive on a particular market. Namely, because of the insufficient competitiveness of most (or some of) domestic farms some countries apply a public protection mode – subsidies, state purchase, price guarantee schemes, border restrictions etc.

2.2. Criteria for Farm Competitiveness

A farm will be competitive if it is *efficient*, and *adaptive*, and *sustainable* [Bachev 2010b]. Thus, there are three *criteria* for assessing the competitiveness of a farm (Figure 1). First, *farm efficiency* – that is the potential of a farm to organize effectively the production *and* transaction activity (of farmer, coalition of members), and minimize the overall production *and* transaction costs. Broadly applied traditional approach cannot assess adequately the efficiency of farms since it restricts analysis to the *technical* efficiency (productivity) and/or *financial* efficiency (profitability). At the same time, significant *transaction costs* associated with the farming organization and farm's potential to economize on governance costs are completely ignored. Farm is not only a production but a *governance* structure as well [Bachev 1996, 2004]. Besides production costs farming activity is usually associated with significant *transaction costs*[1].

For instance, there are costs for studying and complying with various institutional requirements (laws, standards, informal norms); for finding best prices and partners; for identification and protection of diverse property rights; for negotiating conditions of exchange; for contract writing and registration; for setting up and maintaining of a coalition; for enforcing negotiated terms through monitoring, controlling, measuring and safeguarding; for directing and monitoring hired labor; for collective decision making and controlling members of the coalition; for disputing, including through a third party (court system, arbitrage or another way); for adjusting or termination along with the evolving conditions of exchange etc.

---

[1] *Production costs* are the cost associated with proper technology ("combination of production factors") of certain farming, servicing, environmental, community development etc. activity. The *transaction costs* are the costs for governing the economic and other relations between individuals.

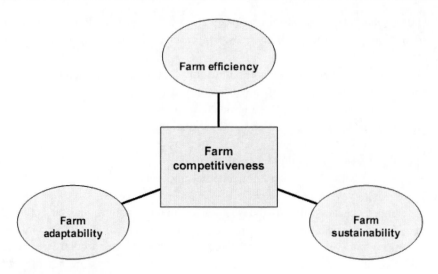

Figure 1. Criteria for assessing competitiveness of farm.

In addition, the choice of type of farming organization is often determined by the *personal characteristics of individual agents* – preferences, ideology, knowledge, capability, training, managerial experience, risk-aversion, reputation, trust, power etc. For instance, if farmer is a good manager he will be able to design and control a bigger organization managing effectively more internal (labor) and outside (market and contract) transactions. A risk-taking farmer will prefer more risky but productive forms - e.g. bank credit for a new profitable venture). When counterparts are family members or close friends there is no need for complex organization since relations are easily "governed" by the good will and mutual interests of parties. Furthermore, *benefits* for farmers could range from monetary or non-monetary income; profit; indirect revenue; pleasure of self-employment or family enterprise; enjoyment in agricultural activities; desire for involvement in environment, biodiversity, or cultural heritage preservation; increased leisure and free time; to other non-economic benefits.

Therefore, the *overall* production *and* transaction costs *and* benefits of a farm are to be taken into account in the assessments of farm efficiency. Different *types* of farms (subsistent, semi-market, part-time, family, group, cooperative, firm, corporative etc.) have *unlike missions, goals, costs* and *benefits* for owners, *modes of enhancement of efficiency* etc. [Bachev 2004]. Therefore, they apply quite different *strategies for development* – e.g. preservation or expansion of a family farm, income support, group farming, servicing members, innovation, commercialization, market domination, specialization, diversification, cooperation with competitors, environmental conservation, integration into processing and food chain, direct (on farm) marketing, international trade etc.

Consequently, diverse farms would have quite *different ways* for expression of their proper efficiency. Thus, it is to be expected a significant variation in the rate of profitability on investments in an agro-firm (a profit-making organization) from the "pay-back" of expenditures or resources in a family farm (a major or supplementary income generation form), in a cooperative (a member oriented organization), in a public farm (a non-for profit organization) or in a semi-market farm (giving opportunity for productive use of otherwise "non-tradable" resources such as family labor, land etc.)[2].

Furthermore, there are many highly effective (non-market, cooperative etc.) farms which are not competitive since they do not compete on market at all. In order to be competitive a farm must be effective *and* be able to govern effectively its *marketing* transactions. Therefore, the system of assessment of farm competitiveness is to take into account the farm's specific *and* market efficiency.

Second, *farm adaptability* – that is farm's potential (ability, incentives) to adapt to constantly changing market, economic, institutional, and natural environment.

A market farm could be very effective (in optimization of current production and transaction costs) but unless it possesses a good adaptation potential it will not be competitive. A market farm must have not only high *historical* or *current* efficiency but a *long-term* ability to perform effectively. The later implies existence of a good potential for farm adaptation to: liberalization of markets, globalization and augmentation of competition; dynamics of demand and prices of farm products; evolution of supply and prices of agrarian inputs, labor, services, finance etc.; progression of public support to farms; development of market and institutional norms, standards and regulations; changes in natural environment (e.g. global warming, extreme weather, water shortages etc.).

For instance, in Bulgaria there are many highly productive (small scale, livestock etc.) farms which are *not able* adapt (lack of managerial ability and/or needed resources) to increasing competitive pressure, and new EU quality, safety, environmental preservation, animal welfare etc. standards, and/or challenges associated with the global climate change [Bachev and Nanseki; Bachev 2010].

There are also marketing farms which have *no incentives* to adapt to new environment. For instance, if a farm/firm is in the end of its life cycle (an old age farmer with no successors) it does not have stimulus for a long-term investment for enhancement of adaptability and competitiveness. Similarly, despite the huge public support for restructuring of so called "semi-market farms" in Bulgaria, the progress in implementation of this measure has been very slow and far behind the targets)because of the lack of interests in beneficiaries.

The farm adaptation is achieved through progressive improvement of the *factors of production* (resources, technologies, varieties of plants and livestock), *production structure* and/or *organization of the farm* (labor organization, internal management structure, management of contractual relations, modernization of organizational form etc.). Thus the system of assessment of farm competitiveness is to take into account the farm's potential for adaptation to specific market, institutional and natural environment.

Third, *farm sustainability* – that is farm's ability to maintain (continue) over time [Bachev 2005; Bachev and Peeters; Bachev 2012a].

A farm could be efficient and adaptive but unsustainable in a medium or long-term. Therefore, such farm is not going to be competitive. For instance, around the world there are many part-time farms which "sustain" during the economic crisis (high unemployment, low income) and "suddenly" disappear once the economic situation improves. Likewise, in

---

[2] Indeed, a significant variation in productivity and profitability has been found in all estimates on "efficiency" of different farms during transition now in countries from Central and East Europe [Bachev, 2004; Csáki and Lerman; Gortona and Davidova; Mathijs and Swinnen; Zawalinska].

western countries there are many unsustainable family farms which managers are in retirement age but there is no successor willing to undertake the enterprise.

Similarly, in Bulgaria there are a great number of otherwise efficient but highly unsustainable in a short to medium-term farms [Bachev 2006, 2010]. Most of these farms are individual or family holding operated by old managers[3], or they are located in mountainous regions and specialized in tobacco production (declining markets, limited alternative employment opportunities), or they are old style production cooperatives (crisis in management, reduction in membership).

Furthermore, a market farm could be inefficient and inadaptable but highly "sustainable", e.g. during transition there were many such farming organizations in Bulgaria (various public farms and firms in the *process* of privatization, reorganization or liquidation). Thus the system of assessment of farm competitiveness is to take into account the farms sustainability in shorter and medium terms along with its efficiency and adaptability.

2.3. Assessment of Farm Competitiveness

The evaluation of the overall competitiveness of an individual farm, or farms of different types, specialization or regions, requires a complex *qualitative* analysis. This assessment is to determine the factors and levels of farm efficiency, adaptability and sustainability in the specific market, economic, institutional and natural environment.

Furthermore, for each criteria one or more *indicators* is to be selected giving idea about (measuring) the level of farm efficiency, adaptability and sustainability.

There are a *great variety* of indicators for evaluating farm's *technical* and *financial efficiency* suggested in textbooks (manuals) and/or practically used by various types of farms in diverse sub-sectors of agriculture and different countries. For assessing farm competitiveness, there is to be selected *few* (key) indicators which best characterize the technical and financial efficiency of the specific type of farm in the conditions of a particular sub-sector, region and country. For instance, for the conditions of Bulgarian market farms the *quantitative* indicators for the levels of labor productivity, land and livestock productivity, profitability of farm, profitability of own capital, liquidity, and financial autonomy, are the most appropriate for evaluation of farm's technical and financial efficiency [Koteva and Bachev] (Table 1). For assessing farm's *governance efficiency* a *qualitative* analysis is needed embracing farm's goals, ownership structure, personal characteristics of the farmer and labor, critical dimensions of different farm transactions, level of internal and outside transaction costs, available governance alternatives; competition, cooperation, integration and/or complementarily with other organizations etc.

Furthermore, according to the farmer's personal preferences, and farm's transacting costs and benefits, it could be found that a particular farm would be highly efficient (or inefficient) with various levels of (combination of the) productivity, profitability, financial security, and financial dependency. For instance, despite the low productivity, profitability and financial independence of many Bulgaria cooperatives, their efficiency for members has been high - non-for profit organization of highly specific for members assets and services with minimum production and/or transaction costs [Bachev 2006].

---

[3] 40% of the farm managers in the country are older than 65 [MAF].

Table 1. Indicators for assessing farm competitiveness

Criteria	Indicators
Farm efficiency	Level of labor productivity
	Level of land and livestock productivity
	Level of profitability of farm
	Level of profitability of own capital
	Level of liquidity
	Level of financial autonomy
	Level of governance efficiency
Farm adaptability	Level of adaptability to market environment
	Level of adaptability to institutional environment
	Level of adaptability to natural environment
Farm sustainability	Level of sustainability

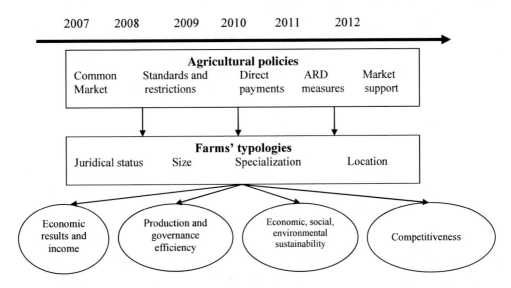

Figure 2. Scope of assessment of CAP impacts on Bulgarian farms.

For assessing *farm's adaptability* three *qualitative* indicators could be used – the level of adaptability to market environment, the level of adaptability to institutional environment, and the level of adaptability to natural environment (Figure 2). Moreover, the level of the *overall adaptability of the farm* will be determined by the indicator with *the lowest* value. For instance, in spite of the high adaptability to market and natural environment of many Bulgarian farms, their overall adaptability has been low since the level of adaptability to the new institutional requirements and restrictions is low [Bachev 2005; Bachev 2010].

For assessing *farm's sustainability* a *qualitative* analysis of the farm and its environment is needed. Some of the factors reducing farm sustainability are *internal* for the farm (e.g. natural "life cycle" of the farm, low efficiency, insufficient adaptability) while others are *external* and associated with the evolution of market, economic, institutional and natural environment.

**Table 2. Identification of type of farm's problems in supply of factors
of production and marketing of output**

Serious problems in:	Character of management problems				
	None	Insignificant	Normal	Big	Unsolvable
Effective supply of needed land and natural resources		☺			
Effective supply of needed labor	☺				
Effective supply of needed material and biological inputs		☺			
Effective supply of needed innovation and know-how			☺		
Effective supply of needed services			☺		
Effective supply of needed funding					☹
Effective utilization and marketing of produce and services				☹	

In order to assess the overall sustainability of a farm a *quantitative* indicator "level of sustainability" could be calculated. *First,* the *managerial problems* associated with the effective supply of needed factors of production and the marketing of output are to be identified, and their *severity* ranged (Table 2). *Persistence* of serious *unsolvable* problems in any of the functional areas of the farm management would indicate a *low governance efficiency* and *sustainability. Next,* the level of sustainability in supply of each of the factors of production and in the marketing of output is to be determined through *transformation* of the "level of problems in management" into the "levels of sustainability" (Table 3). The level of the *overall* sustainability of a farm will coincide with *the lowest* level of sustainability of supply of any of the factors of production or the marketing of products. For instance, despite the high sustainability of supply of natural, human and material factors of production, the overall level of sustainability of most Bulgarian farms is low because of the low sustainability of the management of finance supply and/or marketing of output [Bachev 2005].

In addition to traditional statistical, farming system, and accountancy data, a new type of *micro-economic data* for farm's specific characteristics, activity and governance as well as *data for farm's market, institutional and natural environment* are needed to access the level of competitiveness through various indicators. These *new* data are to be collected through interviews with farm managers and/or experts in the area. The analysis of various aspects of farm competitiveness let not only to determine its level but also to identify the critical factors impeding its improvement, and assist farm management and public policies modernization.

Often, the values of different indicators for individual criteria are with *different directions.* For instance, the efficiency and sustainability of a farm(s) could be high while adaptability low and vice versa. In order to get idea about the *overall* competitiveness of a farm and to be able to make *comparison* of competitiveness of different farms it is necessary to calculate an *Index of Farm Competitiveness.*

Initially, we have to convert the specific value of indicators for efficiency, adaptability and sustainability into universal *unitless* values. An exemplary scale for conversion of the qualitative indicators for overall efficiency, adaptability and sustainability into universal (unitless) indicators is presented in Table 4.

**Table 3. Scale for conversion of levels of management problems
in levels of sustainability**

Seriousness of problems	Level of sustainability
None	Very high
Insignificant	High
Normal	Good
Big	Low
Unsolvable	Unsustainable

**Table 4. Scale for conversion of qualitative indicators for overall efficiency,
adaptability and sustainability into universal indicators**

Qualitative value of indicators			Quantitative value
Level of efficiency	Level of adaptability	Level of sustainability	
Very high	Very high	Very high	1
High	High	High	0,75
Good	Good	Good	0,5
Low	Low	Low	0,25
Insufficient	Insufficient	Insufficient	0

After that, we could calculate an integral Index of Farm Competitiveness (Ic) by multiplying the Index of Farm Efficiency (Ie), Index of farm adaptability (Ia) and Index of Farm Sustainability (Is) using formula: Ic = Ie xIax Is.

The value of Ic would vary between 0 and 1, as a farm would be highly competitive when Ic is 1, uncompetitive when Ic is 0, and with a range of different (low, good etc.) levels of competitiveness when Ic is between 0 and 1. The specific ranges and weights of indicators for assessing farm efficiency and integral competitiveness as high, good, low and insufficient is to be determined by experts according to the specific conditions in each country, subsector of agriculture or type of farming organization.

Depending on identified ranges and weights for assessment, a particular farm would have quite unlike level of the overall competitiveness. For instance, if there is no competition with imported products in a local market, a farm with relatively low productivity will be competitive. On the other hand, the same farm would be uncompetitive in an opened and matured market with a strong internal and international competition.

2.4. Framework for Assessing Impacts of EU CAP on Farms

Introduction of European Union Common Agricultural Policy (EU CAP) in the new member states has profound impact(s) on the competitiveness of farms of different type. Assessment is to be made on the effects on agricultural farms from the implementation of various instruments of EU CAP including (Figure 2):

- common market of agrarian and food products – access to enormous European market, trade liberalization, intensification of competition, common policies toward third countries;

- System of new standards (for quality, hygiene, safety, environmental protection, animal welfare etc.) and restrictions (milk quotas, limits for vineyards extension, reduced use of natural resources etc.);
- Direct payments from EU and national top-ups;
- Support measures of the National Strategic Plan for Agrarian and Rural Development (NPARD);
- Mechanisms of market support of different sub-sectors and exports etc.

The analysis is to embraces effects of CAP implementation on farms as a whole and of different type consisting of:

- farms with different juridical status – physical persons, cooperatives, firms of different type (Sole traders, Limited Liability Companies, Joint Stock Companies, Corporations, Associations etc.);
- farms with different size – rather small, middle size, and rather big for the (sub)sector;
- farms with different specialization–field crops, vegetables, permanent crops, grazing livestock, pigs, poultry and rabbits, mix crops, mix livestock, and mix crop-livestock;
- farms with different geographical locations – predominately plain, predominately mountainous, plain-mountainous, areas with natural handicaps, protected zones and territories.

An assessment is to be made on real rather than "projected or plan" effects of CAP implementation on:

- economics results and income from farms activity;
- change in production and governance efficiency of farms;
- economic, social, and environmental sustainability of farms;
- level of competitiveness of farms.

Assessment is to be based on available official information (public agencies and professional organizations) further resized with original farms survey data and experts evaluations.

3. LEVEL OF COMPETITIVENESS OF BULGARIAN FARMS

3.1. Evolution, Efficiency and Sustainability of Farms

Post-communist privatization of farmland and other agrarian resources has contributed to a rapid development of private farming in the country[4]. There emerged more than 1,7 million private farms of different type after 1990 (Table 5).

---

[4] Agrarian transition was basicaly completed by 2000. Bulgaria joined the EU on Janualy 1, 2007.

Table 5. Evolution and importance of different type farms in Bulgaria

	Public farms	Unregistered	Cooperatives	Agro-firms	Total
Number of farms					
1995	1002	1772000	2623	2200	1777000
2000	232	755300	3125	2275	760700
2005		515300	1525	3704	520529
2010		350900	900	6100	357900
Share in number (%)					
1995		99.7	0.1	0.1	100
2000		99.3	0.4	0.3	100
2005		99.0	0.3	0.7	100
2010		98,0	0,25	1,7	100
Share in farmland (%)					
1995	7.2	43.1	37.8	11.9	100
2000	1.7	19.4	60.6	18.4	100
2005		33.5	32.6	33.8	100
2010		33,5	23,9	42,5	100
Average size (ha)					
1995	338.3	1.3	800	300	2.8
2000	357.7	0.9	709.9	296.7	4.7
2005		1.8	584.1	249.4	5.2
2010		2,9	807	211,6	8,5

Source: National Statistical Institute and Ministry of Agriculture and Food.

Majority of newly evolved farms are *unregistered farms* (Physical persons). They concentrate the main portion of agricultural employment and key productions like livestock, vegetables, fruits, grape etc. (Table 6).Unregistered farms are predominately *subsistence, semi-market* and *small-scale commercial* holdings. According to the official data the farms smaller than 2 European Size Unit (ESU)[5]comprise the major share of all farms in main agricultural subsectors (Figure 3). What is more, in livestock activities they account for the bulk of the Standard Gross Margin (SGM) in related subsectors.

Agrarian reform has turned most households into owners of farmland, livestock, equipment etc. An *internal organization* of available household resources in an own farm has been an effective way to overcome a great institutional and economic uncertainty, protect private rights and benefit from owed resources, and minimize costs of transacting [Bachev 2000].

During transition, market or contract trade of much of household capital (land, labor, money) was either impossible or very expensive due to: unspecified or completely privatized rights, "over-supply" of resources (farmland, unemployed labor), "missing" markets, high uncertainty and risk, asymmetry of information, enormous opportunism in time of hardship, little job opportunities and security etc.

---

[5] 1 ECU=1200 Euro. According to the EU classification farms with a size of 2-4 ESU are considered as "semi-market farms". The actual number of subsistence and semi-subsistence farms is unknown since many of them are not covered by the Agricultural Census.

Table 6. Share of different type farms in all holdings, agrarian resources and productions in Bulgaria

Indicators	Physical persons	Cooperatives	Sole traders	Companies	Associations
Number of holdings with Utilized Agricultural Area (UAA) (%)	99.0	0.3	0.4	0.2	0.05
Utilized agricultural area (%)	30.3	40.3	11.7	16.1	1.6
Average size (ha)	1.4	592.6	118.8	352.5	126.2
Number of breeders without UAA (%)	96.1	0.2	1.9	1.7	0.1
Workforce (%)	95.5	1.2	0.8	1.4	0.3
Labor input (%)	91.1	4.1	1.4	2.8	0.6
Cereals (%)	26.6	41.8	13.0	17.3	1.3
Industrial crops (%)	20.5	45.1	14.2	18.6	1.6
Fresh vegetables (%)	86.4	4.4	4.2	4.6	0.4
Orchards and vineyards (%)	52.3	29.5	2.9	10.7	4.6
Cattle (%)	90.2	5.1	1.5	2.5	0.7
Sheep (%)	96.0	1.4	0.8	1.0	0.8
Pigs (%)	60.3	1.4	7.0	30.5	0.8
Poultry (%)	56.5	0.2	13.3	29.3	0.7

Source: MAF, Agricultural Holdings Census in Bulgaria'2003.

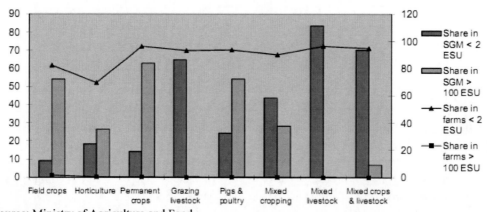

Source: Ministry of Agriculture and Food.

Figure 3. Share of farms with SGM smaller than 2 ESU and bigger than 100 ESU in total SGM and farms with different specialization (percent).

Running up an own farm has been the most effective (or only feasible) mode for productive use of available resources (free labor, land, technological know-how), providing full and part-time employment or favorable occupation for family members, and securing income and effective (cheap, safe, sustainable) food supply for individual households. Specialization or diversification into small-scale farming has taken place [Bachev 2008], and even now the agriculture is an "additional source of income" for one out of 7 Bulgarians [MAF].

Management of the small-scale farms is not associated with significant costs (Table 7).

Table 7. Time and efforts for governing of farm transactions in Bulgaria (% of farms)

Efforts and time for:	Level	Type of farms			Small	Middle	Large	Total
		Unregistered	Cooperative	Firms				
Finding new workers	big	18,91	14,28	12,5	18,91	18,18	0	15,46
	moderate	8,10	42,85	37,5	5,40	45,45	31,25	27,83
Finding partners selling or leasing-out farmland	big	18,91	35,71	12,5	13,51	31,81	12,5	21,64
	moderate	29,72	14,28	62,5	18,91	40,90	62,5	36,08
Finding suppliers for needed materials, equipment etc.	big	24,32	21,42	50	21,62	34,09	50	31,95
	moderate	29,72	67,85	25	35,13	45,45	31,25	39,17
Finding markets for outputs	big	37,83	42,85	56,25	27,02	56,81	56,25	45,36
	moderate	13,51	35,71	28,12	27,02	20,45	31,25	24,74
Finding the rest of needed information	big	45,94	17,85	15,62	40,54	18,18	25	27,83
	moderate	10,81	21,42	40,62	8,10	31,81	37,5	23,71
Negotiating and preparing contracts	big	18,91	35,71	40,62	16,21	40,90	37,5	30,92
	moderate	27,02	21,42	37,5	21,62	27,27	50	28,86
Controlling implementation of contractual terms	big	48,64	42,85	37,5	45,94	36,36	56,25	43,29
	moderate	5,40	14,28	31,25	5,40	22,72	25	16,49
Resolving conflicts associated with quality and contracts	big	29,72	14,28	59,37	29,72	31,81	56,25	35,05
	moderate	5,40	50	21,87	16,21	31,81	18,75	23,71
Relations with banks and preparing projects for crediting	big	35,13	42,85	59,37	32,43	47,72	68,75	45,36
	moderate	8,10	42,85	37,5	5,40	45,45	31,25	16,49
Associating with registration regimes	big	18,91	17,85	15,62	18,91	18,18	12,5	17,52
	moderate	2,70	21,42	9,37	10,81	13,63	0	10,30
Relations with administration	big	24,32	10,71	18,75	21,62	15,90	18,75	18,55
	moderate	21,62	42,85	40,62	32,43	38,63	25	34,02
Relations with membership organizations	big	18,91	21,42	6,25	16,21	20,45	0	15,46
	moderate	5,40	25	43,75	2,70	40,90	25	23,71
Others	big	5,40	14,28	0	0	13,63	0	6,18
	moderate	0	0	0	0	0	0	0

Source: interviews with farm managers[6].

_____

[6] This survey covers 2,8 % of the cooperatives, 1,2 % of the agro-firms, and 0,3% of the unregistered farms in the country as all holdings were selected as representative for the nation's main regions.

Table 8. Share of farms with different level of efficiency in Bulgaria (percent)

Type of farms	Productivity			Profitability			Financial availability			Financial dependency		
	low	good	high	low	good	high	low	good	high	low	average	high
Unregistered	44,83	48,28	6,90	51,72	37,93	10,34	62,07	20,69	17,24	51,72	34,48	13,79
Cooperatives	11,54	84,62	1,92	26,92	73,08	0,00	25,00	75,00	0,00	23,08	53,85	23,08
Firms	11,11	55,56	33,33	33,33	55,56	11,11	33,33	55,56	11,11	22,22	55,56	22,22
Field crops	15,69	74,51	9,80	29,41	64,71	5,88	29,41	60,78	9,804	25,49	54,9	19,61
Mix crop-livestock	38,46	46,15	7,69	46,15	53,85	0,00	46,15	46,15	7,69	46,15	38,46	15,38
Mix crops	33,33	66,67	0,00	50,00	50,00	0,00	41,67	58,33	0,00	33,33	50,00	16,67
Mix livestock	0,00	100,00	0,00	0,00	0,00	100,00	0,00	100,00	0,00	0,00	100,00	0,00
Grazing livestock	100,00	0,00	0,00	100,00	0,00	0,00	100,00	0,00	0,00	100,00	0,00	0,00
Pigs and poultry	100,00	0,00	0,00	100,00	0,00	0,00	100,00	0,00	0,00	100,00	0,00	0,00
Permanent crops	0,00	100,00	0,00	25,00	75,00	0,00	62,50	37,50	0,00	37,5	25,00	37,50
Vegetables	33,33	66,67	0,00	33,33	66,67	0,00	33,33	66,67	0,00	33,33	33,33	33,33
All farms	22,22	70,00	6,67	35,56	60,00	4,44	37,78	55,56	6,67	32,22	47,78	20,00

Source: interviews with farm managers.

Table 9. Share of farms with different level of adaptability in Bulgaria (percent)

Type of farm	Adaptability to:								
	market			institutions			nature		
	low	good	high	low	good	high	low	good	high
Unregistered	51,72	48,28	0,00	31,03	68,97	0,00	37,93	55,17	6,90
Cooperatives	34,62	65,38	0,00	23,08	71,15	5,77	61,54	36,54	0,00
Firms	0,00	66,67	33,33	22,22	22,22	55,56	22,22	44,44	33,33
Field crops	41,18	54,90	3,92	21,57	64,71	13,73	54,90	41,18	3,92
Crop-livestock	38,46	61,54	0,00	38,46	61,54	0,00	38,46	61,54	0,00
Mix crops	25,00	75,00	0,00	16,67	83,33	0,00	58,33	25,00	16,67
Mix livestock	0,00	100,00	0,00	0,00	100,00	0,00	0,00	100,00	0,00
Grazing livestock	100,00	0,00	0,00	0,00	100,00	0,00	0,00	100,00	0,00
Pigs and poultry	100,00	0,00	0,00	0,00	100,00	0,00	0,00	100,00	0,00
Permanent crops	25,00	75,00	0,00	37,50	62,50	0,00	50,00	37,50	0,00
Vegetables	0,00	66,67	33,33	33,33	33,33	33,33	0,00	66,67	33,33
All farms	36,67	60,00	3,33	25,56	65,56	8,89	50,00	43,33	5,56

Source: interviews with farm managers.

Table 10. Share of farms with different level of problems of farm sustainability in Bulgaria (percent)

Type of problems	All farms	Unregistered	Cooperatives	Firms	Field crops	Crop-livestock	Mix crops	Mix livestock	Grazing livestock	Pigs and poultry	Permanent crops	Vegetables
Effective supply of needed land and natural resources												
Insignificant	23,33	37,93	17,31	11,11	23,53	15,38	25,00	0,00	0,00	100,00	25,00	33,33
Normal	61,11	44,83	67,31	77,78	62,75	69,23	66,67	100,00	100,00	0,00	37,50	33,33
Significant	14,44	17,24	13,46	11,11	13,73	15,38	8,33	0,00	0,00	0,00	25,00	33,33
Effective supply of needed labor												
Insignificant	34,44	51,72	26,92	22,22	33,33	30,77	33,33	0,00	0,00	100,00	50,00	33,33
Normal	51,11	31,03	61,54	55,56	50,98	53,85	58,33	100,00	100,00	0,00	50,00	33,33
Significant	14,44	17,24	11,54	22,22	15,69	15,38	8,33	0,00	0,00	0,00	0,00	33,33
Effective supply of needed inputs												
Insignificant	32,22	48,28	25,00	22,22	29,41	46,15	41,67	0,00	100,00	100,00	12,50	0,00
Normal	56,67	31,03	69,23	66,67	66,67	30,77	50,00	100,00	0,00	0,00	62,50	33,33
Significant	11,11	20,69	5,77	11,11	3,92	23,08	8,33	0,00	0,00	0,00	25,00	66,67
Effective supply of needed finance												
Insignificant	30,00	55,17	13,46	44,44	31,37	38,46	25,00	0,00	0,00	100,00	0,00	66,67
Normal	54,44	20,69	73,08	55,56	56,86	30,77	66,67	100,00	0,00	0,00	75,00	33,33
Significant	14,44	24,14	11,54	0,00	9,80	30,77	8,33	0,00	100,00	0,00	25,00	0,00
Effective supply of needed services												
Insignificant	48,89	51,72	44,23	66,67	49,02	46,15	66,67	0,00	0,00	100,00	37,50	33,33
Normal	41,11	27,59	51,92	22,22	43,14	30,77	25,00	100,00	100,00	0,00	62,50	33,33
Significant	10,00	20,69	3,85	11,11	7,84	23,08	8,33	0,00	0,00	0,00	0,00	33,33
Effective supply of needed innovation and know-how												
Insignificant	42,22	62,07	30,77	44,44	43,14	23,08	41,67	0,00	100,00	100,00	50,00	66,67
Normal	36,67	20,69	44,23	44,44	37,25	46,15	41,67	100,00	0,00	0,00	25,00	0,00
Significant	20,00	17,24	23,08	11,11	19,61	30,77	16,67	0,00	0,00	0,00	12,50	33,33
Effective marketing of products and services												
Insignificant	17,78	34,48	5,77	33,33	17,65	15,38	16,67	0,00	100,00	100,00	0,00	33,33
Normal	50,00	37,93	59,62	33,33	56,86	46,15	50,00	100,00	0,00	0,00	12,50	66,67
Significant	30,00	27,59	30,77	33,33	23,53	38,46	33,33	0,00	0,00	0,00	75,00	0,00

Source: interviews with farm managers.

Table 11. Evolution of economic efficiency of Bulgarian farms

Indicators	Field crops			Vegetables			Permanent crops			Grazing livestock			Pigs and poultry		
	2005	2007	2008	2005	2007	2008	2005	2007	2008	2005	2007	2008	2005	2007	2008
Profitability	10,9	33,6	30,6	12,2	8,7	5,64	12,2	8,7	5,64	49,6	42,3	38,07	28,1	12,3	6,91
Land productivity	37	55	78	210	188	253	210	188	253	123	94	109	557	646	466
Labor productivity	9780	17077	21704	14170	11362	14994	14170	11362	14994	4406	6300	7042	7689	10336	7527
Net Income/farm	4273	17467	22432	10295	3780	3733	10295	3780	3733	5484	8284	8759	6920	7251	3606
Net Income/UAA	8	26	34	35	25	22	35	25	22	86	61	66	334	239	116

Source: Ministry of Agriculture and Food.

They are mainly *individual* or *family holdings*, and farm size is exclusively determined by household resources – family labor, own farmland and finance. Internal governing costs are non-existent (one-person farm) or insignificant because the coalition is between family members (common goals, high confidence, and no cheating behavior dominate). Farmers have strong incentives to increase efficiency adapting to internal or market demand, intensifying work, investing in human capital etc. since they own the whole residuals (income).

Nevertheless, there has been a constant decrease in the number of unregistered farms as a result of labor exodus (competition with other farms or industries, retirement, emigration), organizational modernization (change in type of enterprises), increasing market competition (massive failures and take-overs), and impossibility to adapt to new institutional requirements (standards) for safety, quality, environmental preservation, animal welfare etc.

More than 3000 new *production cooperatives* emerged during and after liquidation of ancient "cooperative" structures in 1990s (Table 4). They have been the biggest farms in terms of land management concentrating a major part of cereals, oil and forage crops, and key services to members and rural population (Table 5).

The cooperative has been the *single* effective form for farming organization in the absence of settled rights on main agrarian resources and/or inherited high interdependence of available assets (restituted farmland, acquired individual shares in the actives of old cooperatives, narrow specialization of labor) [Bachev 2000]. After 1990 more than 2 millions Bulgarians have got individual stakes in the assets of liquidated ancient public farms. In addition to their small size, a great part of these shares have been in indivisible assets (large machinery, buildings, processing and irrigation facilities). Therefore, new owners have had no alternative but liquidate (through sales, consumption, distortion) or keep these assets as a joint (cooperative) ownership. In many cases, the ownership rights on farmland were restituted with adjoined fruit trees and vineyards, and much of the activities (e.g. mechanization, plant protection, irrigation) could be practically executed solely in cooperation.

Most "new" landowners happened to live away from rural areas, have other business, be old of age, or possess no skills or capital to start own farms. In the absence of a big demand for farmlands and/or confidence in emerging private farming during first years of transition, more than 40% of the new owners pulled their land and assets in the new production cooperatives.

Moreover, most cooperatives have developed along with the new small-scale and subsistent farming. Namely, "non-for-profit" character and strong member (rather than market) orientation have attracted the membership of many households. In transitional conditions of undeveloped markets, high inflation, and big unemployment, the production cooperative has been perceived as an effective (cheap, stable) form for supply of highly specific to individual farms inputs and services (e.g. production of feed for animals; mechanization of major operations; storage, processing, and marketing of farm output) and/or food for households consumption.

The cooperative rather than other formal collective (e.g. firm) form has been mostly preferred. Cooperatives have been initiated by older generation entrepreneurs and a long-term "cooperative" tradition from the communist period has a role to play. Besides, this mode allows individuals an easy and low costs entree and exit from the coalition, and preservation of full control on a major resource (such as farmland), and "democratic" participation in and control on management ("one member-one vote" principle).

In addition, the cooperative form gives some important tax advantages such as tax exemption on sale transactions with individual members and on received rent in kind. Also for coops there are legal possibilities for organization of transactions not legitimate for other modes such as credit supply, marketing, and lobbying at a nation-wide scale[7].

Relatively bigger operational size gives cooperatives a great opportunity for efficient use of labor (teamwork, internal division and specialization of work), farmland (cultivation in big consolidated plots, effective crop rotation, environment protection), and material assets (exploration of economies of scale and scope on large machinery etc.).

In addition, cooperatives have a superior potential to minimize market uncertainty (dependency) and increase marketing efficiency ("risk pooling", advertisement, storing, integration into processing and direct marketing); and organize some critical transactions (better access to commercial credit and public programs; stronger negotiating positions in input supply and marketing deals; facilitate land consolidation through simultaneous lease-in and lease-out contracts; introduce technological innovations; effective environmental management); and invest in intangible capital (good reputation, own labels, brand names) etc.

In a situation of "missing markets" in rural areas, the cooperative mode is also the single form for organization of certain important activity such as bakery, processing, retail trade, recreation etc.

The cooperative activity is not difficult to manage since internal (members) demand for output and services is known and "marketing" secured ("commissioned") beforehand (Table 6). In addition, cooperatives concentrate on few highly standardized (mass) products (such as wheat, sunflower etc.) with a stable market and high profitability.

Furthermore, the cooperative applies low costs long-term lease for the effective land supply from members. Output-based payment of labor is common which restrict opportunism and minimize internal transaction costs. Besides, cooperatives provide employment for members who otherwise would have no other job opportunities - housewives, pre- and retired persons. Moreover, they are preferable employer since they offer a higher job security, social and pension payments, paid day-offs and annual holidays, opportunity for professional (including career) development. Giving the considerable transacting benefits most cooperative members accept a lower (than market) return on their resources - lower wages, inferior or no rent for land and dividends for shares.

There have been some adjustments in cooperatives size, memberships, and production structure. A small number of coops have moved toward a "business like" (popularly known as "new generation cooperative") governance applying market orientation, profit-making goals, close and small-membership policy, complex joint-ventures with other organizations etc. That has been a result of overtaking the cooperatives management by younger entrepreneurs, improving the governance, taking advantage from new market opportunities and public support programs, and establishing of some of coops as key regional players.

Besides, some cooperatives have benefited significantly from the available new public support (product or area based subsidies), and the comparative advantages to initiate, coordinate and carry out certain (environmental, rural development etc.) projects requiring large collective actions.

---

[7] Forbidden for business firms by the Double-taxation and Antimonopoly Laws.

At the same time, many cooperatives have shown certain *disadvantages* as a form for farm organization. A big membership of the coalition (averaging 240 members per coop) makes individual and collective control on the coop's management very difficult and costly. That gives a great possibility for mismanagement and/or let using cooperatives in the best interests of managers or groups around them (on-job consumption, unprofitable for members' deals, transfer of profit and property, corruption)[8].

What is more, majority of the new cooperatives did not overcome the incentive problems associated with the collective team working in the old public farms - over employment, equalized remuneration, authoritarian management, adverse feeling towards private farming, system of personal plots etc. [Bachev 2006].

Furthermore, there are differences in the investment preferences of diverse members (old-younger; working-non-working; large-small shareholders) due to non-tradable character of cooperative shares (so called "horizon problem"). While working and younger members are interested in long-term investments and growth of salaries, income in kind, other on-job benefits, the older and not working members favor higher current gains (income, land rent, dividend). Given the fact that most cooperative members in the country are small shareholders, and older in (pre-retired and retired) age, and non-permanent employees, the incentives for long-term investment for land improvement and renovation of outdated and physically amortized machinery, buildings, orchards, vineyards etc. have been very low.

Finally, many cooperatives fall short in adapting to diversified (service) needs of members, and evolving market demand and growing competition. For all these reasons, the economic performance of production cooperatives has not been good. Accordingly, the efficiency of cooperatives has diminished considerably in relation to other modes of organization (market, contract, partnership etc.). Many landlords have pooled out their land from the cooperatives since property rights on farmland were definitely restored in 2000. Consequently, a significant reduction of cooperative activity has taken place and a big amount of cooperatives ceased to exist in recent years.

There has been a "boom" in creation of different type *agri-firms* after 1990 as their number and importance have augmented enormously (Table 5). They account for a tinny portion of all farms but concentrate a significant part of UAA, material assets, major productions and significant portion of the SGM of cereals, industrial crops, orchards, poultry and swine (Table 6, Figure 3).

Business farms are commonly *large specialized enterprises*. Most of them have been set up as *family* and *partnership* organization during first years of transition by younger generation entrepreneurs - former managers (specialists) of public farms, individuals with high business spirit and know-how etc. Majority of these farms are formally registered as *Sole Traders*. In addition, some state farms and agri-firms have been taken over by former managers and teams and registered as *Shareholdings* (Companies, Associations). Furthermore, different sort of *joint ventures* with non-agrarian and foreign capital increasingly appear as well.

The specific management skills and the "social" status as well as the combination and complementarities of partner's assets (technological knowledge, business and other ties, available resources) have let a rapid extension of business farms through enormous

---

[8] The latter has been "assisted" by the lack of any (outside) public control on the cooperative's activity.

concentration of (management of, ownership on) resources, and exploration of economies of scale and scope, and modernization of enterprises [Bachev 2000].

The specific mode and the pace of privatization of agrarian resources have facilitated a fast consolidation of the fragmented land ownership and agrarian assets in the large farms. During the long period of institutional and market transformation (unsettled rights on resources, imperfect regulations, huge uncertainty and instability) the personal relations and "quasi" or entirely integrated modes have been extensively used to overcome transaction difficulties.

Furthermore, the large operational size of these enterprises gives enormous possibilities to explore technological opportunities (consolidation of land, economies of scale and scope on machineries, cheap and standardized produce etc.) and achieve a high productivity. Business farms have been constantly extending their share in managed agrarian (and related) resources taking over smaller farms, incorporating new types of activities, and applying new organizational schemes.

Business farms are strongly *market* and *profit-oriented* organizations. Farmer(s) have great incentives to adapt to market demand and institutional restrictions investing in farm specific (human, material, intangible) capital because they are sole owners of residual rights (benefits). The owners are commonly family members or close partners, and the internal transaction costs for coordination, decision making, and motivation are not high (Table 7). Increased number of the coalition (partnership) gives additional opportunity for internal division of labor and profiting from specialization – e.g. full-time engagement in production management, technological development, market and "public" relations, paper works, keeping up with changes in laws and standards etc.

Their large size and reputation make business farms a preferable partner in inputs supply and marketing deals. Besides, these farms have a giant negotiating power and effective (economic, political) mechanisms to dominate markets and enforce contracts. They also possess a great potential to collect market and regulatory information, search best partners, promote products, adjust to new market demand and institutional requirements, use outside experts, prepare business and public projects, meet formal (quantity, quality, collateral) requirements, "arrange" public support, bear risk and costs of failures.

In addition, business farms effectively explore economies of scale and scope on production *and* management - e.g. "package" arrangement of outside funding for many projects; interlinking inputs supply with know-how supply, crediting, marketing etc.

Furthermore, large farms have strong incentives and potential for innovation – available resources to test, adapt, buy, and introduce new methods, technologies, varieties; possibility to hire leading (national, international) experts and arrange direct supply from consulting companies or research institutes. What is more, they are able to invest a considerable relation-specific capital (information, expertise, reputation, lobbying, bribing) for dealing with funding institutions, agrarian bureaucracy, and market agents at national or even at international scale. The last but not least important, these farms have enormous political power to lobby for Government support in their best interests. All these features give considerable comparative advantages of business type of farming organization.

The *firm mode* is increasingly preferred since it provides considerable opportunities:

- to overcome coalition difficulties - e.g. formation of joint ventures with outside capital, dispute ownerships right through a court system etc;

- to diversify into farm related and independent businesses - trade, agro-tourism, processing, etc;
- to develop firm-specific intangible capital (advertisement, reputation, brand names, public confidence) and its exploration (extension into daughter company), trade (sell, licensing), and intergeneration transfer (inheriting);
- to overcome existing institutional restrictions - e.g. for direct foreign investments in farmland, trade with cereals, vine and dairy etc.;
- to have explicit rights for taking parts in particular types of transactions - e.g. export licensing, privatization deals, assistance programs etc.

3.2. Level of Competitiveness of Commercial Farms

The assessment on the competitiveness of commercial farms in the country has found out that the majority of surveyed farms[9] are with a *good* and *high* competitiveness (Figure 4). Nevertheless, more than a fifth of all farms are with a *low* level of competitiveness.

Furthermore, different types and kinds of farms are with *unequal* competitiveness. Diverse *agri-firms* (Sole traders and Companies) are with good competitive positions and the portion of enterprises with high competitiveness is particularly big. On the other hand, a quarter of *cooperatives* are with insufficient competiveness.

Most of the highly competitive farms are specialized in *mix livestock*[10] and *vegetables*. For all other groups of specialization, the farms with a good competitiveness comprise the greatest share in respective groups. In *mix crop-livestock*, *mix crops* and *permanent crops* every forth farm is non-competitive.

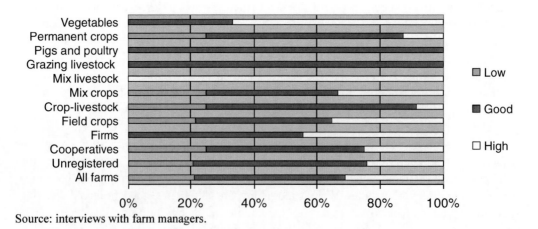

Source: interviews with farm managers.

Figure 4. Share of farms with different levels of competitiveness in Bulgaria.

---

[9] Assessment of competitiveness is based on 2010 interviews with farm managers of 58 unregistered holdings, 104 cooperatives, and 18 agri-firms from all regions of the country.

[10] The number of surveyed farms in groups with specialization in "Mix livestock", "Grazing livestock", and "Pigs and poultry" is very small (only 2).

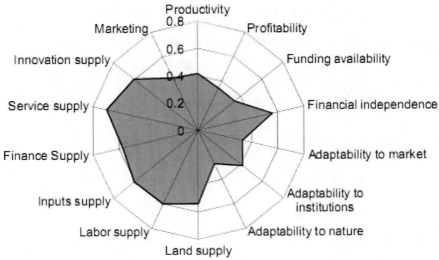

Source: interviews with farm managers.

Figure 5. Importance of individual elements of farm competitiveness in Bulgaria.

The analysis of different *aspects* of the farms competitiveness shows that the farms' low productivity, profitability and funding availability, and insufficient adaptability to market, institutional and natural environment, and serious problems in financial and innovation supply and in marketing of products and services, all contribute to the greatest extend to decreasing the overall level of farms competitiveness (Figure 5).

The analysis of the *level of efficiency* of diverse type of farms shows that majority of farms have a good productivity, profitability, financial availability and financial independence (Table 8).

However, according to the managers of a considerable number of unregistered holdings, and grazing livestock, pigs and poultry, and mix crop-livestock farms the *productivity* of their farms is low.

Furthermore, *profitability* of 36% of all farms is evaluated as low, and more than a half of unregistered farms, and a considerable fraction of mix crop-livestock, mix crops, grazing livestock, and pigs and poultry farms are in this group.

A significant portion of farm managers declare that *availability of finance* is insufficient, and unregistered holdings, farms specialized in mix crop-livestock, mix crops, grazing livestock, pigs and poultry, and permanent crops, suffer the most from the lack of funding. Only a fifth of survey farms are heavily *dependent from outside funding* (credit, state support etc.) as share of highly dependent farms specialized in permanent crops and vegetables is the greatest.

The analysis of the *level of adaptability* of surveyed farms has found out that more than a quarter of them are with a low potential for adaptation to *new state and EU quality, safety, environmental etc. standards,* almost 37% are less adaptable to *market demand, prices and competition,* and every other one is inadaptable to *evolving natural environment* (warning, extreme weather, droughts, floods, etc.) (Table 9).

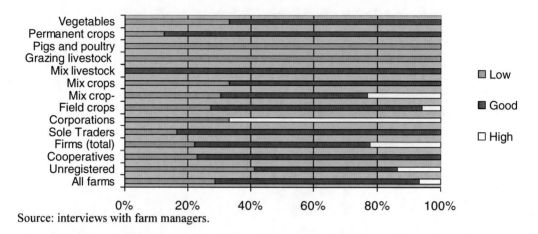

Source: interviews with farm managers.

Figure 6. Share of farms with different levels of medium-term sustainability in Bulgaria.

As far as *farm medium-term sustainability* is concerned, it is evaluated by 29% of the farms managers as low. The share of unregistered holdings, grazing livestock, and pigs and poultry farms with a small sustainability is the biggest (Figure 6).

On the other hand, less that 7% of all farms "forecast" a high mid-term sustainability. A particular type of firms – the *companies,* is the only exception among surveyed farms, and two-third of these enterprises envisages being highly sustainable in years to come.

Detailed analysis of the diverse *factors* diminishing farms long-term efficiency and sustainability indicates that the *significant problems* in the effective *marketing of products and services,* and in the effective *supply of needed innovation and know-how,* are the most important for the good part of surveyed farms (Table 10). Apparently, the later farms have no (internal) adaptation potential to overcome these type of problems and will be unsustainable (inefficient) is a longer run[11].

The serious (unsolvable) problems associated with the *marketing* are critical for a considerable section of agri-firms, and farms specialized in mix crop-livestock, and permanent crops. The severe problems in the effective *supply of needed innovation and know-how* are most important for the sustainability of cooperatives, mix crop-livestock, and vegetable farms.

Furthermore, great difficulties ineffective *supply of needed land and natural resources* face a quarter of farm specialized in vegetables and permanent crops. Harsh problems in effective *supply of needed labor* are critical only for grazing livestock holdings.

Big difficulties in effective *supply of needed inputs* experience a good fraction of unregistered holdings, and farms specialized in vegetables, permanent crops, and mix crop-livestock production. Significant problems in effective *supply of needed finance* are reported by a main part of unregistered holdings, and farms specialized in grazing livestock, mix crop-livestock, and permanent crops. Finally, substantial difficulties in effective *supply of needed services* are common for a big section of unregistered holdings, and farms specialized in permanent crops and mix crop-livestock operations.

---

[11] These farms either have to restructure production, or reorganize farm (new governance), or will disappear in near future.

3.3. Competitiveness of Different Type of Farms

The majority of surveyed *unregistered holdings* are with a *good* level of competitiveness, and around 24% of them are *highly* competitive (Figure 7). At the same time, more than a fifth of all unregistered farms are not competitive.

Unregistered holdings with a different specialization are with *unequal* competitiveness. Most highly competitive farms are in *vegetables, field crops*, and *mix livestock* productions. On the other hand, a half of the holdings in *permanent crops*, a third of all farms in *mix crops* and 29% of *mix crop-livestock* operators are with a low level of competitiveness.

The analysis of different *components* of the competitiveness of unregistered holdings indicates that the low productivity, profitability, and funding availability, along with the insufficient adaptability to changing market, institutional and nature environment, and the severe problems associated with marketing of products, are mostly responsible for diminishing the competitiveness of these farms (Figure 8).

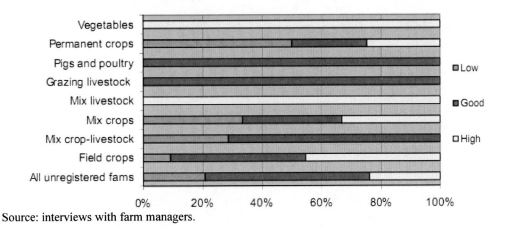

Source: interviews with farm managers.

Figure 7. Share of unregistered farms with different levels of competitiveness in Bulgaria.

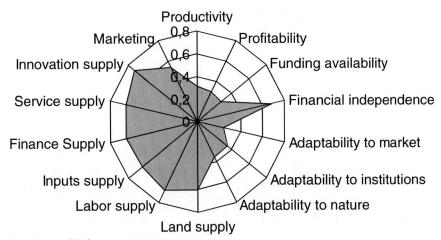

Source: interviews with farm managers.

Figure 8. Importance of individual elements of competitiveness of unregistered farms in Bulgaria.

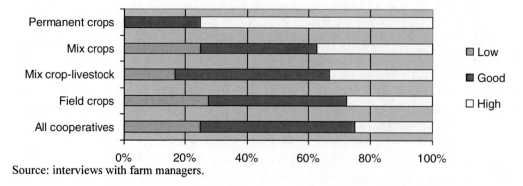

Source: interviews with farm managers.

Figure 9. Share of cooperatives with different levels of competitiveness in Bulgaria.

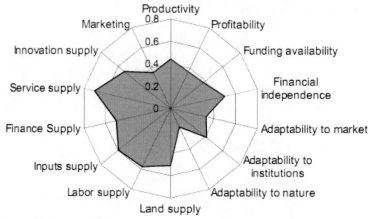

Source: interviews with farm managers.

Figure 10. Importance of individual elements of competitiveness of cooperatives in Bulgaria.

On the other hand, the higher efficiency in supply of factors of production and the lower dependency from outside funding, enhance the overall competitiveness of unregistered farms.

A half of surveyed *cooperatives* are with a *good* level of competitiveness, and a quarter of them are *highly* competitive (Figure 9). At the same time, one out of four cooperatives is not competitive. The cooperatives with a diverse specialization are with *different* level of competitiveness. Most of the highly competitive cooperatives are in *permanent crops* and *mix crops*. At the same time, a significant number of cooperatives in *field crops* and *mix crops* are with a low level of competitiveness. The analysis of different *elements* of the competitiveness of cooperatives shows that the low productivity, profitability, financial availability and independency, together with the insufficient adaptability to market, institutional and nature environment, and the difficulties associated with finance, land and innovation supply and marketing mainly affect the reduction of competitiveness of cooperatives (Figure 10). All surveyed *agri-firms* are with a *good* or a *high* competitiveness. What is more, a significant number of these farms (44%) are highly competitive (Figure 11). Nevertheless, while three-quarter of the firms in *field crops* are with high level of competitiveness, all firms in *mix crops* and *permanent crops* are with a good competitiveness, and *vegetables* producers are equally divided in good and high competitive groups.

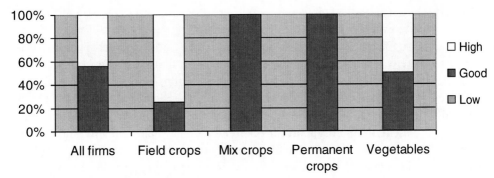

Source: interviews with farm managers.

Figure 11. Share of agri-firms with different levels of competitiveness in Bulgaria.

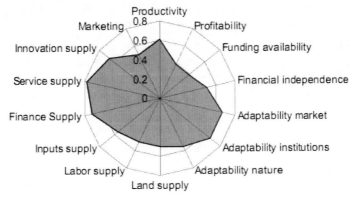

Source: interviews with farm managers.

Figure 12. Importance of individual elements of competitiveness of agri-firms in Bulgaria.

The analysis of individual *factors* the competitiveness of agri-firms exposed that the low productivity, profitability, funding availability and independency, and the serious problems in labor and land supply and marketing, greatly contribute to decreasing firms competitiveness (Figure 12). On the other hand, the high adaptability of firms to evolving market and institutional environment, and their considerable efficiency in finance, innovation and service supply raise the overall competitiveness of these farming enterprises.

4. IMPACTS OF EU CAP ON INCOME, EFFICIENCY, SUSTAINABILITY AND COMPETITIVENESS OF BULGARIAN FARMS

4.1. CAP Effects on Farms Economic Results and Income

According to the *experts*[12] the overall impact from implementation of the various CAP mechanisms (common market, market intervention, new standards, direct payments, support

---

[12] Expertise was carried out in the end of 2011 with the 13 leading experts on farm structure and policies in Bulgaria.

from NPARD, export subsidies) on *incomes* of different type of farms is multidirectional. The majority of experts estimate that CAP effect on income of cooperatives, firms, middle and large size farms, and farms specialized in field crops is *good* or *significant* (Figure 13). What is more, most experts evaluate CAP impact on middle size farms and cooperatives rather as *good*, while that on firms and big farms is rather *significant*. Namely larger farming organizations (such as agri-firms and cooperatives) highly specialized in certain field crops (wheat, sunflower, corn etc.) have benefited the most from the major CAP instruments for income and farm modernization support (direct area-based payments, NPARD measures) due to the large farmlands under management, high capability to apply for public support etc. Having in mind the relatively low-income level in many farms (e.g. producers cooperatives) during pre-accession period, it could be concluded that CAP implementation has been associated with a "sizeable" improvement in farms income in the country.

On the other hand, the biggest part of the experts assesses as *insignificant* the impact of CAP on unregistered farms, small holdings, and farms specialized in vegetables, permanent crops, and mix livestock. Furthermore, a good part of the experts estimate as *neutral* or even *negative* the CAP effect on small farms, and holdings specialized in vegetables, permanent crops, grazing livestock, pigs, poultry and rabbits, mix crops, and mix crop-livestock farms.

The majority of surveyed *farm managers*[13] assess as *good* or *significant* the overall impact from implementation of diverse CAP instruments on the *economic results* of their own farms (Figure 14).

All questioned cooperatives, farms with big size, holdings specialized in crop-livestock, pig, poultry and rabbits, and those located in regions with natural handicaps, report a high positive impact from the implementation of the common policy of the Union.

The effect of CAP implementation on economic results is the most *significant* for the surveyed farms with big sizes, cooperatives, specialized in mix crop-livestock and field crops, and situated in the plan and plan-mountainous regions of the country.

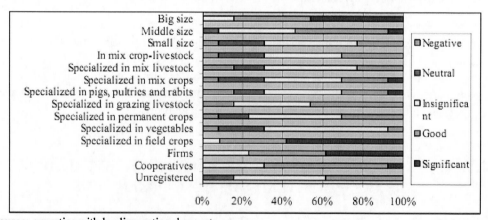

Source: expertise with leading national experts.

Figure 13. Impact of EU CAP on income of Bulgarian farms.

---

[13] A survey with 84 managers of "representative" commercial farms of all type of juridical status, sizes, specialisations, and geografical locations was conducted in the spring of 2012. The structure of surveyed farms approximately correspond to the current structure of comercial farms in the country.

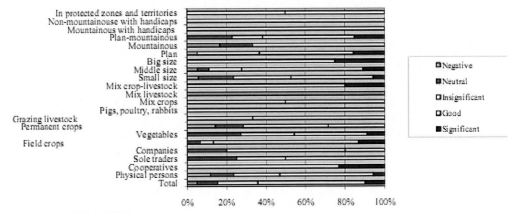

Source: interviews with farm managers.

Figure 14. Impact of EU CAP on farms economic results in Bulgaria.

The weakest positive impact of CAP is on economic results of firms and unregistered farms, holdings specialized in mix livestock and crops, permanent crops and vegetables, and farms with small size, and with lands in protected areas and territories. What is more, all farms with mix livestock, a considerable section of farms in mountainous regions, with permanent crops and Physical Persons, and a portion of farms with small size and in plan regions, estimate as *negative* the impact of the new policy on their economic results.

Implementation of different mechanisms of CAP affects also positively the *incomes* of a great part of surveyed farms. The effect on income is strongest for large farms, cooperatives, farms specialized in mix crop-livestock, pigs, poultry, and rabbits, and field crops and holdings located in plan regions and in areas with natural handicaps. Moreover, 40% of crop-livestock farms, every forth of big farms, and a good part of cooperatives and firms mainly from mountainous regions, assess as *significant* the CAP impact on their income.

Nevertheless, for a considerable fraction of questioned farms CAP implementation is not connected with a positive impact on incomes. The effect on income growth is weakest for farms specialized in mix livestock and crops, permanent crops, grazing livestock and vegetables, firms and unregistered holdings, small farms and farms in mountainous and plan-mountainous regions. For a good part of farms in permanent crops and a portion of unregistered holdings, and farms with middle size in plan regions of the country, the effects of the new policy on income is even *negative*.

Available data also proves that the bulk of public subsidies go to a few number of large farms (agri-firms and cooperatives) specialized in field crops. At the same time, many effective small-scale farms receive no or only a tiny fraction of the public support. For instance, despite it increased number only 24% of all farms received area based direct payments, and merely 6% of cattle holdings, 4% of sheep and pig holdings, and 3% of poultry farms [MAF]. Moreover, less than 7% of the beneficiaries get the lion share (more than 80%) of direct payments. Similarly, due to restrictive criteria, unattainable formal requirements, high costs for participation, and widespread mismanagement (and corruption) the new public support under NPARD is not effectively utilized and benefits a small portion of the farms [Bachev 2010b]. All these further foster the income disparity in different type of farms.

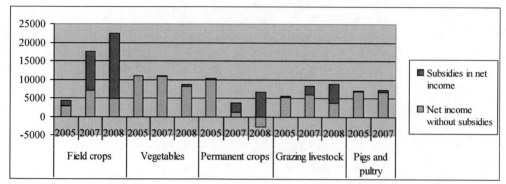

Source: MAF, Agro-statistics.

Figure 15. Evolution of income and public support of different type of Bulgarian farms.

Nevertheless, CAP subsidies are becoming an important part of the net income of farms specialised in filed crops, permanent crops and grazing livestock (Figure 15). Furthermore, subsidies accounts for the major and increasing part of the net income of large farms – 89% (42% in 2007) and 83% (75% in 2007) for farms with 8-40 ESU and above 40 ESU accordingly [MAF].

4.2. CAP Effects on Farms Efficiency

The overall impact of EU CAP on the production efficiency of farms of different types is also unequal. According to the majority of experts the effects of CAP on production efficiency of middle sized holdings and cooperatives is good (Figure 16). The impact on firms, big size farms, and farms specialized in field crops, is estimated as good or significant. In the past years many farms have been improving their efficiency through progressive change in organization, technology, production structure, and introduction of innovation, taking advantage from the new opportunities of public support, market demands etc. On the other hand, most experts assess as *insignificant* the effect of CAP on production efficiency of unregistered farms, and holdings with mix livestock, mix crops, and mix crop-livestock. For the rest type of holdings, the impact of CAP is evaluated as *insignificant, neutral* or even *negative* in relation to production efficiency of farms.

According to the half of the *managers* of surveyed farms the CAP implementation is affected *good* or *significantly* production efficiency (Figure 17). The positive impact of the new policy is strongest on the production efficiency of cooperatives; farms specialized in mix crop-livestock and field crops, farms with big sizes, in plan-mountainous regions, regions with natural handicaps, and in protected areas and territories. Also the main part of surveyed Physical Persons, holdings specialized in vegetables and grazing livestock, farms with small and middle sizes, and those in predominately plan regions, evaluate as *good* or *significant* the effect of new policy on their production efficiency.

Nonetheless, at the same time CAP implementation is having no positive impact on production efficiency of all or a major portion of firms, holdings in permanent crops, poultry and rabbits, and mix livestock, and farms in predominately mountainous regions and in protected zones and territories.

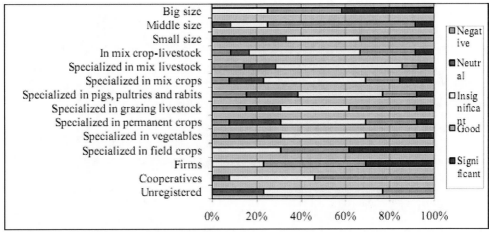

Source: expertise with leading national experts.

Figure 16. Impact of EU CAP on production efficiency of Bulgarian farms.

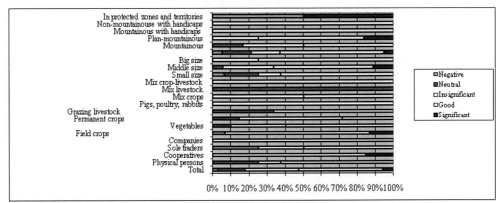

Source: interviews with farm managers.

Figure 17. Impact of EU CAP on farms production efficiency in Bulgaria.

Furthermore, implementation of the new policy is associated with *negative* results in relation to the production efficiency of every other farm with plots in protected zones and territories, a quarter of Sole Traders, and a portion of farms in files crops, small size, and in plan regions of the country.

Dynamics of the main indicators of economic efficiency also demonstrate that there is a positive impact of CAP implementation on profitability, land and labour productivity, and income per farm and utilized land of farms specialised in filed corps (Table 11). For farms specialised in vegetables, permanent crops, and livestock, the evolution of production efficiency indicators is rather negative.

The overall impact of CAP on *governance efficiency* of farms is also quite diverse. The biggest number of *experts* estimate that the overall impact of CAP implementation on the governance efficiency of large farms and the farms specialized in field crops is *good* (Figure 18). For the middle size farms that impact is defined as *insignificant* or *good*.

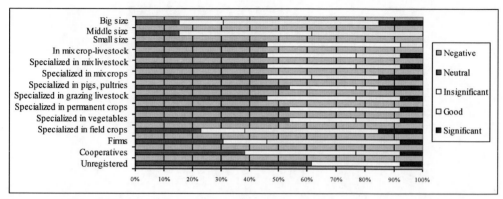

Source: expertise with leading national experts.

Figure 18. Impact of EU CAP on governance efficiency of Bulgarian farms.

Most expects assess the CAP effect on governance efficiency of unregistered holdings, and farms specialized in vegetables, permanent crops, and pigs, poultry and rabbits as *neutral*, and for the rest type of farms as *neutral* or *insignificant*.

Our survey also proves that impact of CAP on governance efficiency of specific types of farms is quite different. More than 47% of the *managers* assess as *good* or *significant* the effect on their governance efficiency, including all of the holdings in regions with natural handicaps, more than 83% of cooperatives, above 69% of farms in field crops, two-third of farms with grazing livestock, and 60 and more percent of holdings with middle sizes, farms specialized in vegetables, and those located in mountainous regions.

The effect in relation to improvement of managerial efficiency is particularly strong for the farms in protected zones and territories where every another one evaluate as *significant* the impact of the new policy. The CAP implementation affects particularly strongly the governance efficiency of a good share of cooperatives, and farms in filed crops, middle sizes, and in plan-mountainous regions of the country.

On the other hand, CAP implementation contributes *insignificantly* or *neutrally* to governing efficiency of all or a major part of farms specialized in pigs, poultry and rabbits, Sole Traders, Physical Persons, Companies, large farms, and holdings specialized in permanent crops and mix production, and those in plan and plan-mountainous regions, and with areas in protected zones and territories. What is more, all farms with mix livestock, one fifth of holdings in predominately mountainous regions and companies, and a good portion of farms in permanent crops, in plan-mountainous regions, with smaller sizes, and Physical Persons, report a *negative* effect of CAP on their governance efficiency. Changes in the market and institutional environment associated with the CAP introduction (enhanced competition; high quality, safety, environmental etc. standards; available public support) affect the internal comparative and absolute potential of the principle type of farming organisations to economise on transaction costs and benefit from the adaptation to the evolving socio-economic environment. Moreover, a number of CAP measures aim at enhancing (certain aspects of) managerial efficiency of (certain type of) farms – e.g. "Semi-subsistence farming", "Setting up producer groups", "Provision of farm advisory and extension services", public eco-contracts etc. Nevertheless, the progress of implementation of specific measures has been slow while the number of affected farms insignificant [Bachev

2012b].Similarly to the past, mostly bigger farms participate in the public support programs because they have a superior managerial and entrepreneurial experience, available resources, possibilities for adaptation to the new requirements for quality and other standards, potential for preparing and wining projects, etc.

Therefore, CAP support measures benefit exclusively the largest structures and the richest regions of the country, and do not contribute to decreasing economic and eco-discrepancy between farms, sectors, and regions.

4.3. CAP Effect on Sustainability of Farms

According to the most *experts* the impact of CAP implementation on economic, social and environmental sustainability of large farms, firms, and farms specialized in field crops is *good* or *significant* (Figure 19). The overall effect of CAP on sustainability of other type of farms is estimated as *insignificant* or *neutral*.

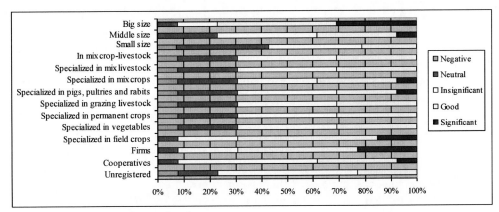

Source: expertise with leading national experts.

Figure 19. Impact of EU CAP on economic, social and environmental sustainability of Bulgarian farms.

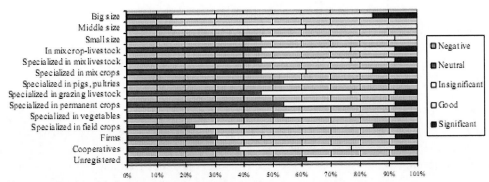

Source: interviews with farm managers.

Figure 20. Impact of EU CAP on farms economic sustainability in Bulgaria.

According to the managers CAP implementation is having good or significant effect on economic sustainability of more than a half of surveyed farms (Figure 20). To the greatest extent the new policy leads to enhancing economic sustainability of cooperatives, big and middle size farms, holdings specialized in mix crop-livestock and filed crops, and farms located in regions with natural handicaps and plans. The impact of CAP is particularly beneficial for increasing the economic sustainability of farms with crop-livestock specialization, of large farms and cooperatives, where the effects is evaluated as significant by each third, each forth and almost 17% of them accordingly. For a part of farms in plan-mountainous regions, in field crops, with middle sizes, the effect on improvement of economic sustainability is also sensible.

On the other hand, for all or a major part of farms in pig, poultry and rabbits, companies, Physical Persons, specialized win permanent crops and grazing livestock, holdings with small sizes and in mountainous regions, the impact of CAP implementation in *insignificant* or *neutral* in relations to economic sustainability.

What is more, all farms specialized in mix livestock, every another one of holdings in mountainous regions and in protected zones and territories, a quarter of Sole Traders, a fifth of companies, and a good fraction of Physical Persons, small and middle size holdings, farms specialized in permanent crops, vegetables and filed crops, and those located in mainly plan regions, assess as diminishing *(negative)* the effects of the new policy on their economic sustainability.

More than a half of surveyed farms also indicate a *good* or *significant* impact of CAP on *social sustainability* of farms, including each tenth one significant effect for improving social sustainability. Implementation of CAP instruments has a favorable impact on social sustainability of all cooperatives (including for almost 17% of them *significant*), all holdings in regions with natural handicaps, every four out of five farms with mix crop-livestock specialization (including for one fifth of them in a *significant* extent), two-third of farms in predominately mountainous regions (including for almost 17% of them *significant*), more than 64% of farms in filed crops (including for more than 7% *significant*), and above 61% of holdings with middle sizes (including for almost 17% of them *significant*). CAP implementation enhances the social sustainability of the half of farms in mix crops (all in *significant* extent), of farms situated in plan regions of the country (including for more than 11% *significantly*), and of the farms with large sizes and in the protected zones and territories.

The CAP contribution to the social sustainability is smallest for mix livestock farms, holdings with pigs, poultry, and rabbits, firms of all type, and farms specialized in permanent crops and grazing livestock. Moreover, CAP implementation is associated with diminishing *(negative* effect) of social sustainability of a portion of surveyed farms –accordingly for more than 14% of specialized in permanent crops, and almost 6% of Physical Persons and farms in plan regions of the country.

As far as impact of CAP on *environmental sustainability* of farms is concerned for more than a half of surveyed holdings it is positive, mostly evaluated as *good* by managers. The favorable effect of CAP on eco-sustainability is felt by all farms with areas with natural handicaps, forth-fifth of holdings in vegetables and mountainous regions, three-quarters of farms in crop-livestock production, more than two-third of farms with grazing livestock, more than 69% of farms in plan-mountainous regions, 60% of Physical Persons, more than 58% of cooperative, and every other farm with small and muddle sizes, in field crops, mix crops, and pigs, poultry and rabbits.

The surveyed the farms do not report a negative impact of CAP on environmental aspects of their activity. Nevertheless, for all holdings with mix livestock and with areas in protected zones and territories, and the majority of farms with permanent crops, plan regions, and big sizes, the effect from implementation of CAP instruments on environmental sustainability is *insignificant* and/or *neutral*.

CAP implementation tends to improve the eco-performance of commercial farms. There is "eco-conditionality" for participating in public programs. In addition, direct payments are inducing farming on previously abandoned lands, and improve eco-situation. Furthermore, there is huge budget allocated for special eco-measures and the number of farms joining agri-environmental programs gradually increases.

CAP measures affect positively the environmental sustainability particularly of large business farms and cooperatives. These enterprises are under constant administrative control (and punishment) for obeying new eco-standards, strongly interested in transforming activities according to new eco-norms (making eco-investments, changing production structures), and realizing economies of scale and scope from participation in special agro-environmental measures.

On the other hand, many small and (semi) subsistence holding can hardly meet new eco-standards and stay in the gray and informal sector. The latter is particularly true for numerous livestock holdings most of which do not still comply with the new EU standards for quality, safety, animal welfare and eco-performance.

4.4. CAP Effect of Farms Competitiveness

Most *experts* assess the overall impact of CAP on the competitiveness of firms, big size farms, and holdings specialized in field crops as *good* and *significant* (Figure 21). The effect on the competitiveness of middle size farms, and holdings specialized in vegetables is determined as *insignificant* or *good*.

According to the *managers* of 42% of surveyed farms the CAP implementation is having a *good* or *significant* impact on their own competitiveness (Figure 22).

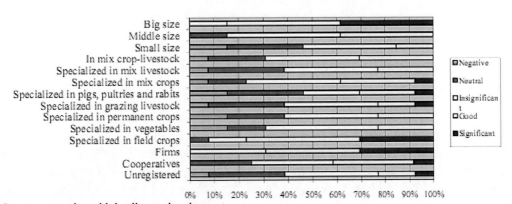

Source: expertise with leading national experts.

Figure 21. Impact of EU CAP on competitiveness of Bulgarian farms.

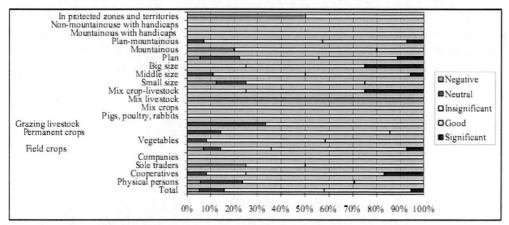

Source: interviews with farm managers.

Figure 22. Impact of EU CAP on farms competitiveness in Bulgaria.

To the greatest extent the new policy improves the competitiveness of holdings in regions with natural handicaps, big and middle size farms, cooperatives and Sole Traders, and farms specialized in mix crop-livestock operations, field crops and grazing livestock. What is more, a quarter of large farms and those with crop-livestock specialization, nearly 17% of the cooperatives, more than 11% of the holdings in predominately plan regions, above 7% of the holdings in field crops and in plan-mountainous regions, and almost 6% of the holdings with middle sizes, estimate as *significant* the effect from CAP implementation for increasing their competitiveness.

On the other hand, all or a main part of the companies, Physical Persons, farms with mix crops, pigs, poultry and rabbits, permanent crops and vegetables, and holdings located in mountainous and plan-mountainous regions, describe as *insignificant* or *neutral* the impact of the new policy on the level of their competitiveness. For all farms with mix livestock, for every other one in protected zones and territories, for one quarter of Sole Traders, for a fifth in predominately mountainous regions, for more than 12% of small holders, for 7% of farms in field crops, and nearly for 6% of Physical Persons and farms in plan regions, the CAP implementation decreases their competitiveness (reported *negative* effect).

4.5. Dynamics of Main Farms Indicator Comparing to the Period before EU CAP Implementation (End of 2006)

The greatest share of surveyed farms indicates an increased level of a part of the main indicators in the present time comparing to the levels in the period before EU CAP implementation (Figure 23). For instance, *higher* or *considerable higher* is the level of the total income, costs, investments, profit, labor productivity, efficiency of the production and management in the majority of surveyed farms. Also the biggest portion of holdings has an improved access to public support, and augmented amount of subsidies for production, income and investment support. At the same time, the share of farms with *lower* total indebtedness comparing to the pre-accession period is 38%, while with a *higher* one below 18%.

Over a half of the farms say that they have improved qualification and information, agro-techniques and crop rotation, and livestock conditions, as well as increased product and food safety, and innovation activity compared to the period before CAP implementation. All of which is a direct or indirect result of the favorable impact of different CAP mechanisms on the key aspects of the activities of majority of surveyed farms.

However, a good fraction of farms report *lack of change* in share of sold output, market access, diversification of products and services, deepening of specialization, and in environmental preservation. Also a big part of farms have no changes in their dependency from suppliers and buyers, increased integration with suppliers and buyers, and improved involvement in professional organizations and access to the agricultural advisory system.

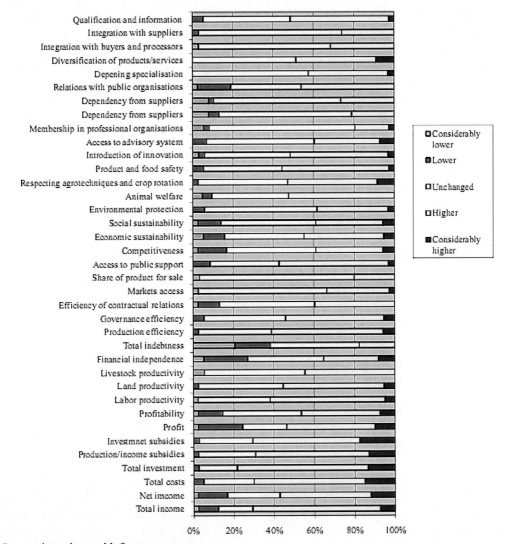

Source: interviews with farm managers.

Figure 23. Level of farms major indicators comparing to level before EU CAP implementation in Bulgaria.

Furthermore, a big portion of holdings do not report changes in the profitability, land and livestock productivity, overall indebtedness and financial independency, efficiency of production, management and contractual relations, competiveness, economic and social sustainability, agro-techniques and crop rotation, livestock conditions, product and food safety, introduction of innovation, qualification and information. Besides, more than a third of farms have no improvement in the relations with state organizations and in the access to public support in comparison to the pre-accession period.

Therefore, implementation of diverse instruments of CAP does not lead to a progressive change in the man indicators of a good part of farms. The latter is either due to the lack of positive effect from CAP on a portion of holdings (for example, lack of effective public support) or due to neutralized effect of CAP on other negative factors which could have deteriorated even further the state of farms (in conditions of lack of counterbalancing the existing negative trends CAP instruments).

For a considerable share of farms, the current level of the main indicators is *lower* or *significantly lower* compared to the level before CAP introduction. For instance, 27% of surveyed holdings indicate deteriorated financial independence, more than 24% are with diminished profit, almost 17% are with reduced net income and competitiveness, around 16% are with inferior economic sustainability, almost 15% are with lower profitability, and 14% are with deteriorated social sustainability. Similarly, nearly 19% of farms are with worsened relations with the state organizations, above 13% of them have decreased efficiency of contractual relations, every tenth is with inferior livestock conditions, almost 9% of holdings are with decreased access to public support, and more than 8% are with reduced membership in professional organizations.

All these show that CAP implementation is associated with deterioration of main indicators of a considerable portion of farms. This is either because of the negative effects of CAP on a party of farms, or due to the lack of effective mechanisms for assisting the farms adaptation and for compensating the influence of other negative factors (e.g. competition with heavily subsidized imported products at the national and international markets, high interest rates of bank credits, big market price fluctuations etc.).

Figure 24 illustrates the extent and the directions in which the main farms indicators have been changed during the period of CAP implementation in the country. Implementation of diverse CAP mechanisms is associated with significant progressive changes in some of the aspects of activity of a relatively big share of farms. For other aspects of farms activity the CAP implementation does not lead to sensible effective change in the majority of holdings. What is more, in certain directions the effect of CAP is negative for a good portion of farms.

All these necessitate improvement of the CAP implementation through perfection of management public programs, change in design and/or beneficiaries of some CAP instruments, or require rethinking and reforming individual mechanisms or the policy as a whole.

According to the managers the CAP implementation affects quite unlikely the competitiveness of different type of farms. As a result of improved market and institutional environment and public support, and increased investment and efficiency of farms, the competitiveness of two-third of surveyed farms *increases*, including for each fifth one is a *significant scale* (Figure 25).

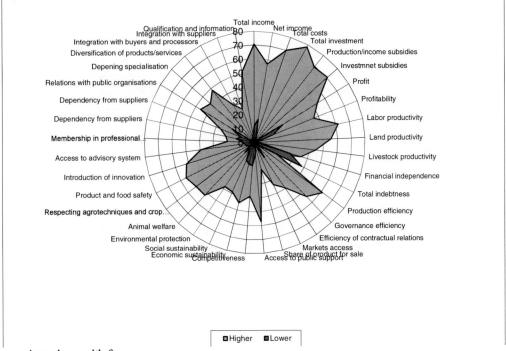

Source: interviews with farm managers.

Figure 24. Dynamics of main farm indicators comparing to the pre-accession period in Bulgaria (percent of farms).

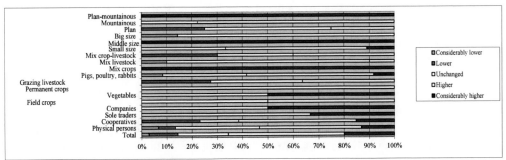

Source: interviews with farm managers.

Figure 25. Current level of farms competitiveness comparing to the pre-accession period in Bulgaria.

During the period of CAP implementation the competitiveness *increases* of all type of firms, holding specialized in mix livestock and vegetables, and farms located in plan regions and in protected zones and territories. The majority of cooperatives, farms with big sizes, mix crops, and non-mountainous areas with natural handicaps also record a growth in competitiveness.

Nevertheless, CAP implementation the country is *not* associated with a change in the competitiveness of farms specialized in grazing livestock, main part of small holdings, and farms in plan-mountainous regions and in mountainous areas with natural handicaps, and a good portion of Physical Persons, cooperatives, farms in field crops, pigs, poultry and rabbits, mix crops, middle and large size holdings. Moreover, the current level of the competitiveness of 30% of middle sized farms, more than 27% of holdings specialized in pigs, poultry and rabbits, a quarter of farms in the mountainous areas with natural handicaps, more than 23% of cooperatives, above 14% of farms in plan-mountainous regions, more than 13% of Physical Persons, every tenth of smallholdings, and more than 8% of mix crop farms, is *lower* or *significantly lower* comparing to the period before CAP introduction.

Therefore, CAP implementation does not contribute to improvement of competitiveness of a great portion of farms in the country.

CONCLUSION

We have demonstrated that the New Institutional and Transaction Costs Economics is a powerful methodology which let us better understand the "logic" and adequately assess the farm efficiency and competitiveness in the specific market, institutional and natural environment of Bulgarian agriculture.

The analysis of the post-communist transition and EU integration of Bulgarian agriculture has found out that fundamental property rights and institutional modernization has been associated with the evolution of a specific farming structure consisting of numerous small-scale and subsistent holdings and a few large cooperatives and agro-firms. Furthermore, agrarian agents have developed and use a great variety of effective contractual arrangements to govern their relations, resources and activities – formal, informal, simple, complex, interlinked, market, private, collective, bilateral, trilateral, multilateral, hybrid etc.

Various type of farms have quite different efficiency, adaptability, and sustainability in the specific Bulgarian conditions of undeveloped markets, badly defined and/or enforced formal rights and rules, inefficient forms of public intervention, specific "Bulgarian" way of EU "common" policies implementation, dominant informal "rules of the game" etc. What is more, diverse farming organizations possess unlike competitive advantages in rapidly changing market, institutional and natural environment. While most market farms are with a good competitiveness, a great part of agri-firms are highly competitive, and a considerable fraction of unregistered holdings and cooperatives uncompetitive.

EU CAP implementation in the country affects in dissimilar ways the income, efficiency, sustainability and competitiveness of farms of different types. It has got an overall positive impact on cooperatives, firms of different type, big farms, holdings specialized in field crops, and farms located in plan regions and areas with natural handicaps. Despite that the CAP implementation affects favorably the income, efficiency, sustainability and competitiveness of a portion of other type of holdings, the overall impact of CAP for the majority of agricultural holdings in the country is either insignificant or neutral. What is more, for a good fraction of small holdings, unregistered farms, farms specialized in vegetables, permanent crops, livestock, and mix crop-livestock, and holdings in mountainous regions the CAP implementation has been associated with negative effects.

REFERENCES

Bachev, H. (1996). *Organization of Agrarian Transactions in Transitional Economies*, paper presented at the 8th Congress of the European Association of Agricultural Economists "Redefining the Roles for European Agriculture", 3-7 September, Edinburgh.

Bachev, H. (2000). Bulgarian Experience in Transformation of Farm Structures, *Farm Management and Rural Planning No 1*, Fukuoka: Kyushu University Press, 181-196.

Bachev, H. (2004). Efficiency of Agrarian Organizations, *Farm Management and Rural Planning No 5*, Fukuoka: Kyushu University Press, 135-150.

Bachev H. (2005). *Assessment of Sustainability of Bulgarian Farms*, paper prepared for presentation at the XIth Congress of the EAAE "The Future of Rural Europe in the Global Agri-Food System", Copenhagen, Denmark, August 24-27, 2005 www.eaae2005.dk/POSTER_PAPERS/ SS34_16_Bachev.pdf

Bachev, H. (2006). Governing of Bulgarian Farms – Modes, Efficiency, Impact of EU Accession, In: J., Curtiss, A., Balmann, K. Dautzenberg, and K. Happe, (editors), *Agriculture in the Face of Changing Markets, Institutions and Policies: Challenges and Strategies*, Halle (Saale): IAMO, 133-149.

Bachev, H. (2007). National Policies Related to Farming Structures and Sustainability in Bulgaria, In: A., Cristoiu, T., Ratinger, S. Gomez, and Y. Paloma, (Editors), *Sustainability of the Farming Systems: Global Issues, Modeling Approaches and Policy Implications*, Seville: EU JRC IPTS, 177-196.

Bachev, H. (2008). Post-Communist Transformation in Bulgaria-Implications for Development of Agricultural Specialization and Farming Structures, in S. Ghosh (Editor), *Agricultural Transformation: Concepts and Country Perspectives*, Punjagutta: The Icfai University Press, 91-115.

Bachev, H. (2010a). *Governance of Agrarian Sustainability*, New York: Nova Science.

Bachev, H. (2010b). *Management of Farm Contracts and Competitiveness*, Saarbrucken: VDM Verlag.

Bachev, H. (2012a). Framework for Evaluating Sustainability of Farms and Agrarian Organisations, *International Journal of Social Sciences and Interdiciplinary Research*, October Issue.

Bachev, H. (2012b). Modes, Challenges and Opportunities for Risk Management in Modern Agri-food Chains, *The IUP Journal of Supply Chain Management*, Vol.IX, No 3, 24-51.

Bachev, H. and Tsuji, M. (2001). Structures for Organization of Transactions in Bulgarian Agriculture, *Journal of the Faculty of Agriculture of Kyushu University, No 46(1)*, 123-151.

Bachev, H. and Kagatsume, M. (2002). Governing of Financial Supply in Bulgarian Farms, *The Natural Resource Economics Review No 8*, Kyoto: Kyoto University Press, 131-150.

Bachev, H. and Manolov I. (2007). *Inclusion of small scale dairy farms in the supply chain in Bulgaria (a case study from the Plovdiv region)*, Regoverning Markets Innovative Practice series, London: International Institute for Environment and Development.

Bachev, H. and Nanseki, T. (2008). *Risk Governance in Bulgarian Dairy Farming*, paper presented at the 12th Congress of the European Association of Agricultural Economists "People, Food and Environments–Global Trends and European Strategies", 26-29 August 2008, Ghent, (http://ageconsearch.umn.edu/bitstream/44136/2/240.pdf).

Bachev, H. and Peeters, A. (2005). Framework for Assessing Sustainability of Farms, *Farm Management and Rural Planning No 6*, Fukuoka: Kyushu University, 221-239.

Benson, G. (2007). *Competitiveness of NC Dairy Farms*, North Carolina State University, http://www.ag-econ.ncsu.edu/faculty/benson/DFPPNatComp01.PDF

Csáki, C. and Lerman, Z. (2000). *Structural change in the farming sectors in Central and Eastern Europe*, World Bank Technical Paper Volume 465, Washington DC.

Fertő, I. and Hubbard, L. (2001). *Revealed Comparative Advantage and Competitiveness in Hungarian Agri-food Sectors*, 2002KTK/IE Discussion Papers 2002/8, Institute of Economics Hungarian Academy of Sciences, Budapest. http://econ.core.hu/doc/dp /dp/mtdp0208.pdf

Gortona M. and Davidova, S. (2003). *Farm productivity and efficiency in the CEE applicant countries: a synthesis of results*, Elsevier B.V..

Koteva, N. and Bachev, H. (2011). A Study on Competitiveness of Bulgarian Farms, *Economic Tought*, 7, 95-123.

Mathijs, E. and Swinnen, J. (1997). *Production Organization and Efficiency during Transition: An Empirical Analysis of East German Agriculture*, Policy Research Group, Working Paper No. 7, http://www.agr.kuleuven.ac.be/aee/clo/prgwp/prg-wp07.pdf

MAF (2009).*Agrarian paper.* Sofia: Ministry of Agriculture and Food.

Mahmood, K., Saha, A., Gracia, O. and Hemme, T. (2004).*International competitiveness of small scale dairy farms in India/Pakistan*, http://www.tropentag.de/2004/abstracts /full/376.pdf

NSI (2009).*Statistical Book.* Sofia: National Statistical Institute.

North, D. (1990). *Institutions, Institutional Change and Economic Performance.* Cambridge: Cambridge University Press.

Popovic, R., Knezevic, M. and Tosin, M. (2009).*State and Perspectives in Competitiveness of one farm type in Serbia*, paper presented at the 113 EAAE Seminar "The Role of Knowledge, Innovation and Human Capital in Multifunctional Agriculture and Territorial Rural Development", Belgrade, December 9-11, 2009. http://ageconsearch.umn.edu /bitstream/ 57416/2/Popovic%20Rade%20cover.pdf

Pouliquen, A. (2001). *Competitiveness and farm incomes in the CEEC agri-food sectors. Implications before and after accession for EU markets and policies*, http://ec.europa.eu /agriculture/publi/reports/ceeccomp/sum_ en.pdf

Shoemaker, D., Eastridge, M., Breece, D., Woodruff, J., Rader, D. and Marrison, D. (2009).*15 Measures of Dairy Farm Competitiveness*, http://ohioline.osu.edu/b864 /pdf/864.pdf

Zawalinska, K. (2005). *Changes in Competitiveness of Farm Sector in Candidate Countries Prior to the EU Accession: The Case of Poland*, Paper presented at the 11th Congress of the EAAE "The Future of Rural Europe in the Global Agri-Food System". Copenhagen, August 24-27, 2005. http://ageconsearch.umn.edu/bitstream/24520/1 /cp05za01.pdf.

In: Agricultural Research Updates. Volume 5
Editors: P. Gorawala and S. Mandhatri

ISBN: 978-1-62618-723-8
© 2013 Nova Science Publishers, Inc.

Chapter 2

PERENNIAL WEEDS IN ARGENTINEAN CROP SYSTEMS: BIOLOGICAL AND ECOLOGICAL CHARACTERISTICS AND BASIS FOR A RATIONAL WEED MANAGEMENT

Marcos E. Yanniccari[1,2] and Horacio A. Acciaresi[3,4]

[1]Plant Physiology Institute (INFIVE), National Research Council of Argentina
CONICET– CCT La Plata, La Plata, Buenos Aires, Argentina
[2]Eco-physiology of weeds, Facultad de Ciencias Agrarias y Forestales, Universidad Nacional de La Plata, La Plata, Buenos Aires, Argentina
[3]Scientific Research Commission of the Province of Buenos Aires CICBA, Buenos Aires, Argentina
[4]Plant Production Deparment, Eco-physiology of weeds, Facultad de Ciencias Agrarias y Forestales, Universidad Nacional de La Plata, La Plata, Buenos Aires, Argentina

ABSTRACT

In the last two decades, cropland changes from conventional tilled systems to zero tillage systems have been detected in Pampean agroecosystems. Dominant species are the primary weed due to the fact that they are adapted to the cropping system. Weed population shifts were observed when conventional tillage systems were changed to non-tillage. However, a great proportion of perennial species would be expected in non-till environments. In Argentinean crop systems only a few perennials have transcended in importance.

In this context, *Sorghum halepense* has been one of the most important species in dispersion and aggressiveness. In addition, the low sensitivity of several populations to glyphosate contributes to their complex management. Populations from diverse ecological regions have differential mechanisms for adaptation according to the different environments where they have been growing.

Even in non-till systems, *Cynodon dactylon* is another summer crops primary perennial weed and it is considered as highly plastic species in response to growth factors. Water and radiation have been major factors affecting the growth, conditioning the aggressiveness of the weed.

Recently, the combining uses of zero tillage and round-up ready soybeans have promoted an increased shift in weed herbicide resistance. The detection of a glyphosate-resistant *Lolium perenne* population has increased interest in researchers of these biotypes. Several physiological traits influence their management in winter cereal crops.

The plasticity of these weeds to adapt to zero tillage systems requires a design of integrated weed management program according to biological traits of the weeds. Several strategies of mapping and monitoring of weed populations, crop rotations, biocontrol and rotation of herbicides, among other, could be used to maintain the competitive ability of crops rather than eradication of weeds.

INTRODUCTION

Weeds are successful because they have evolved in response to cropping system practices by adapting and occupying niches left available in agroecosystems [1]. Weed communities have changed over time in response to control practices imposed upon them [2]. The spontaneous vegetations have always been a problem in agricultural crop production and changing agricultural practices have modified and will continue to modify the weed flora [3]. Tillage and other production practices, environmental conditions, herbicide applications, and mechanical weed control practices combined to create the selection pressures that regulate the weed species composition [4].

Weed Communities as Affected by Cropland Changes

In Argentina, a significant agriculture transformation was motorized by the adoption of transgenic crops (soybean, maize, and cotton) under the non-tillage system during the 1990s [5]. The innovation of zero-till farming had far-reaching effects on Argentine agriculture and beyond [6]. As in other countries, non-till cropping has been used to reduce soil erosion, improve soil physical structure, conserve soil water, restore organic matter, and reduce energy consumption [7, 8]. Changes in tillage, crop rotation, herbicide use patterns and other management practices can affect weed species diversity and population demographics [9]. Puricelli and Tuesca [10] evaluated the effect of the regular and exclusive use of glyphosate in crop sequences on associated weed composition, richness and diversity. They concluded that the effect of glyphosate application was the main factor to explain weed community changes in summer crops.

At the beginning of widespread adoption of the non-tillage system, changes in the agroecosystem caused by land use and tillage practices were reflected in the soybean weed community from the Argentine rolling Pampa [11]. Floristic composition differed from 1995 to 2003 and there were more annuals (35 species) than perennials (23 species). Thus, some species that dominated soybean weed community in the past maintained their constancy values; others increased or decreased them at different rates. Other woody species, such as *Fraxinus Americana* L., *Gleditsia triacanthos* L. and *Sida rhombifolia* L. that were only observed in waste and non arable lands in the past, became new components of cropland communities. Permanent non-tillage favored the establishment of these species, which may avoid herbicide damage or may recover from its effects [12]. In other comparisons, 23 years later, the genus *Lolium spp.* has maintained its constancy values above 50% in croplands the south of Buenos Aires Province [13].

Other species that presented high constancy values in the past, such as *Sorghum halepense* (L.) Pers., dominated weed community only at the beginning and decreased their constancy in 2003 [12]. But in 2007, *Sorghum halepense* increased its importance when the first detection of a glyphosate-resistant population from north-west of Argentina was communicated by Vila-Aiub *et al.* [14]. Since then, several glyphosate-resistant populations were detected along the northern and central areas of Argentina [15].

In the west of Buenos Aires Province, Argentina, where no tilled cropping system is predominantly used, the area invaded with *Cynodon dactylon* (L.) Pers. has been increasing in the last decade [16]. The stolon fragments as vegetative propagules of *Cynodon dactylon* are main organs for the expansion of the weed in non-tillage cropping systems [17]. The combination of tillage and glyphosate is more effective in the control of *Cynodon dactylon* than these elements applied separately [18]. In this way, despite the mentioned changes in the Pampean agroecosystems, this weed is still a threat.

The evolution of individuals in existence, the extent of their reproductive success of any organism is to be measured in terms of numbers output, the area of the world's surface that they occupy, the range of habitats that they could occupy, and their potentiality for putting their descendants in a position to continue the genetic line through time [19]. In this sense, only a few perennials species (herbaceous) have been transcendent as problematic weeds in Argentinean crop systems. In this chapter, we have reviewed relevant information related to the eco-physiology of perennial weeds as *Sorghum halepense*, *Cynodon dactylon* and *Lolium perenne*. Finally, the information shown is related to give some prospect in the development of integrated crop–weed management strategies, leading to maintain the competitive ability of crops rather than eradication of weeds.

Classification and Biological Characteristics of Perennial Weeds

Weeds could be classified in many ways and with various aims. In this sense, the taxonomic classification is an important botanical grouping; however, even species that are closely in this system of classification may differ greatly in characteristics of ecological importance. Hakansson [20] suggested a classification of weeds using an ecological sense. Thus, the perennial plants are grouped in: *monocarpic perennials*: plant individuals need more than two growing seasons to reach their generative phase and the plant normally dies entirely following the formation of generative reproduction organs; and *policarpic plants*: the weed survives winter or dry seasons by means of vegetative structures and can repeat vegetative development and growth and form organs for generative reproduction in more than one year. In the last, two sub-groups are distinguished: *stationary perennials* of slight horizontal extension of individuals or clones through shoots or roots (the normal mode of reproduction is seed), and *creeping perennials* with horizontal extension of individuals and clones through plagiotropic shoots or roots.

The reproductive strategies mentioned above are related to the predominant type of reproduction of each species in an ecological environment. Thus, the reproduction by seed is an important mechanism of infestation of a new area and this organ promotes the survival in unfavorable environments, often for long time. The ability to reproduce vegetatively or asexually through rhizomes, stolons, bulbs, corms, aerial bulblets or tubers is related to the capacity of horizontal extension of clones. And, usually, these organs are carbohydrates reserve sinks that help to perpetuate the species. In this way, perennial weeds are usually more difficult to control than annuals because they persist and spread by vegetative organs as well as seeds.

Traits of Perennial Weeds of Importance in the Crop-weed Interaction

Plant interactions may involve positive and neutral or negative (*i.e.*, competition) effects on the performance of plants sharing limited resources [21]. A better competitor will suffer, in turn, less competition from its neighbors. Thus, enhanced competitive ability may be explained by an increased resource capture or, vice versa, by an increased tolerance to deficits of resources [22]. Under this view, several traits of perennials weeds contribute to its competitive ability:

- High rate of early growth: When the dormancy is broken, after the unfavorable season, the use carbohydrates from the reserve organs allow a rapid early growth and the expansion of a leaf canopy to intercept light. Furthermore, there is a fast root penetration for the uptake of water and nutrients.
- Multiplication capacity during the life cycle: Creeping perennials reproduce by vegetative structures producing new individuals (clone) in a growth season.
- Re-growth capacity: The elimination of perennial weeds requires the killing of all plant parts that are capable of producing new shoots. The control of aerial organs does not ensure the eradication of the weed, often vigorous re-growth from underground organs become ineffective weed control techniques.

SORGHUM HALEPENSE IN ARGENTINA: HISTORY AND PERSPECTIVES FROM BIOLOGY ASPECTS

Description and Life Cycle

Sorghum halepense (Johnsongrass) is a monocot weed in the Poaceae family, having a C4 photosynthetic pathway. This was indigenous to the Mediterranean region from the Madeira Islands to 'Asia Minor' [23]. It was introduced to South America (Argentina and Uruguay) as a summer forage grass and later it was extended as an important weed over large areas [24].

Johnsongrass is a perennial plant up to 1.50 m tall, with culms arising from an extensively creeping and rooting scaly rhizome. The rhizomes are long, of horizontal or oblique growth to deepen the soil profile. The leaf blades are 20-40 cm of length and 1-2 cm of width. Panicles are mostly 20-40 cm long. The spikelets of 5-7 mm of length, ovate, shortly awned are in pairs, one spikelet being, hermaphrodite, sessile and the other masculine and pedicellate. Terminal spikelets group are in triplets of one sessile and two pedicellate spikelets. Sessile spikelets distinctly articulate with the subtending joint and are readily deciduous with the accompanying pedicellate spikelet and superior joint [24, 25].

This species is spread from seeds and rhizomes due to invasive characteristics, it is considered one of the world's worst weeds and was declared national weed in Argentina since 1930. The pattern of growth and development of Johnsongrass seedlings and rhizome sprouts are similar, although in the field rhizome sprouts are thought to emerge earlier and to grow more rapidly than seedlings [26].

Johnsongrass is mainly an autogamous plant, but 6 to 8% of seed produced is result of cross-pollination. In Argentina, the production of seed has been reported around 40,000 seed.m^{-2} in areas of high infestation density [27]. The growth from seeds is the principal mode of infestation of new areas. Seed dormancy is important in Sorghum halepense because it maintains a reservoir of seed bank in the soil. Recently harvested Johnsongrass seeds are dormant, however it varies with ecotype. Seed dormancy mechanism lies in the sheath of the caryopsis and consists of a mechanical resistance to the growth of the embryo [28].

Seed germination required a temperature of 20°C. Approximately, 1 week after emergence of the plants a rhizome spur is evident. After formation of this spur, rhizome growth is relatively slow and subordinate to the much more rapid top growth. At 3 weeks after emergence seed, stalks started to emerge from the newly formed crowns at the base of the plants. This species has a 12 to 13 h photoperiod. In these conditions, flowering commences and continues for the rest of the growing season. After the start of blooming, rhizome production increases rapidly. The growth from rhizomes originated from axillary buds occurs at temperatures higher than 15°C, and the optimum for rhizome sprouting is 28°C. These organs have no real dormancy period, but temperature and moisture conditions affect the sprout of rhizomes [26, 29, 30]. While herbicides and cultivations may eradicate young seedlings, control means against plants with rhizomes are much more difficult because they must achieve complete killing of the rhizomatous system in order to be effective [31].

Once established, rhizomes are the main way by which the plant expands. This weed produces three kinds of rhizomes, called primary, secondary and tertiary. Primary rhizomes overwinter and sprout the following spring. Three to five weeks after sprouting, they usually die or lose the capacity to produce new sprouts. As the primary rhizomes buds, it produces secondary rhizome which begins to develop aerial stems as soon as it reaches the surfaces of the soil, thus forming the top. The secondary rhizomes and base and crown of the new plant produce new rhizomes which are also called secondary rhizomes when they produce new aerial stems or clumps of stems. Newly-produced rhizomes which tend to move deeper into the soil and do not sprout during the same growing cycle but overwinter are called tertiary rhizomes. These tertiary rhizomes become the primary rhizomes of the following year [32].

Underground organs growth is positively correlated with root exudation. The allelopathic effects of *Sorghum halepense* and identification of bioactive allelochemicals have been well documented, and much information exists on identification of bioactive allelochemicals in this weed and those exuded from living roots. Johnsongrass has yielded high levels of root exudates per fresh root weight compared to other species of *Sorghum* genus. Theses exudates contain four to six major metabolites, with sorgoleone as the major component (> 85%) [33].

Sorgoleone presence has resulted in inhibition of shoot growth. Thus, a concentration-dependent inhibition of growth was observed in several species. Nimbal *et al.* [34] have determined that sorgoleone inhibited photosystem II electron transport. The inhibitory effects of *Sorghum halepense* on crops may be vary in different Johnsongrass ecotypes [35]. In conclusion, this allelochemical could be an important metabolite in the interference weed-crops.

Plasticity and Response to Soil Water Availability

Johnsongrass causes direct and indirect losses mainly in maize, soybean, grain sorghum and sunflower. Several studies have shown that crop yield losses varied between 12 to 95% in corn, 19 to 99% in sunflower, and 18 to 94% in soybean for low and high Johnsongrass infestation levels, respectively [32, 36, 37]. This weed interferes in crop production through competition and allelopathy, this latter was mentioned above.

The competitive ability of *Sorghum halepense* in agricultural systems of the Argentinean Pampas, as well as several ecophysiological and demographical features of this species, has been evaluated in different studies. Competition for water is one of the most important processes that cause water stress, resulting in reductions in both weed growth and crop yield. As soil water availability decreased, *Sorghum halepense* ecotypes have shown a higher competitive ability, maintaining tolerance to competition and increasing their suppressive capacity to the crop. Under low soil water potential, the stomatal conductance, transpiration rate and photosynthetic rate have been maintained in Johnsongrass. This response is in agreement with the findings of Patterson [38] and seems to be a general response of many weeds, which supports the view of weeds as 'water wasters'.

Sorghum halepense populations from humid and sub-humid regions have shown differential physiologic characteristics. The ecotypes collected in humid regions have had higher relative growth rate when grown at field capacity. As soil water availability decrease, populations from sub-humid regions exhibit a higher relative growth rate and root length ratio than those from humid regions. These differences between ecotypes might be related to the dynamics of the root system: root length relative vary according to the ecotype and soil water availability. Under field capacity, Johnsongrass from the humid region have had lower root length ratio (less proportion of root biomass) as compared to plants from the sub-humid region. Plants from sub-humid region also showed a significant increase in the length of roots per biomass unit, and in the total root biomass under drought. This 'root plasticity' seems to be a developed trait of the 'sub-humid' ecotypes [39]. The adjustments in biomass and also in root length may have allowed the maintenance of water absorption, gas exchange and growth under low soil water availability. The ecotypes from a humid region made a smaller root adjustment, but it still allow these plants to maintain active gas exchange and growth.

The decrease in soil water availability causes morphological (*e.g.*, root mass fraction) and physiological (*e.g.*, gas exchange) changes in *Sorghum halepense*. Heschel *et al.* [40] determined that the magnitude of the plastic responses may vary among the populations of the same species, which may promote both types of adjustments (i.e., morphological and physiological), thus contributing to a reproductive homeostasis. Sexton *et al.* [41] also determined that a generalist and a specialist strategy may not be mutually exclusive, as they may occur to different degrees according to the different stages of the crop-weed interaction. This view may be widely applied to increase the knowledge on the adaptive changes that weeds could perform as the agroecosystem suddenly changes, as it has recently occurred in the extensive crop production systems of Argentina [39].

In interaction with maize under water competition, *Sorghum halepense* has been less affected by the competition than that the crop. The weed was able to maintain its ability to grow, due to the maintenance of net assimilation rate during competition. A similar contribution of biomass from leaves and rhizomes to roots was observed in Johnsongrass, favoring the formation of very fine roots. In contrast, in maize, the decrease in root growth rate was due to a decline in net assimilation rate and there was no formation of fine roots to maintain water absorption during competition. The greater increase both in biomass partitioned to roots and root length by *Sorghum halepense*, might negatively impact maize ability to compete for water during the critical competition period [42].

Johnsongrass and Attempts at Control Using Herbicides

As discussed above, this species was introduced in Argentina a century ago as forage crop, and still invades a wide variety of agro-ecosystems, despite the tremendous efforts to control it [24]. Traditional control of Johnsongrass has included fallow plowing or disking, as well as crop rotation and mowing [43]. The first attempts to control it using herbicides were made by using sodium chlorate, but the main use of chemical products were MSMA (monosodium salt of methylarsonic acid) and DSMA (disodium methanearsonate) combined with appropriate tills during decade 1960's [24].

Up to 1973, there were no selective methods which would make it possible to grow summer crops on heavily infested fields. After this year, a selective control method was developed for rhizomes and seed generated *Sorghum halepense* in soybean using double-dose trifluralin and the split-application technique. These included mainly mefluidide with post-emergence application on soybean [44] and pyriphenop sodium on sunflower, potatoes, cotton and other broadleaf crops [32].

The introduction of aryloxyphenoxy propionate and cyclohexanedione herbicides in the 1980's provided highly effective postemergence control of Johnsongrass in many broadleaf crops [45, 46]. These groups and the sulfonylurea chemical family had reduced the Johnsongrass frequency in crop fields by the mid-90s [47]. However, *Sorghum halepense* ecotypes from humid and sub-humid regions have shown different sensitivity to nicosulfuron, but the control commercial field rates have been acceptable [48].

Tassara *et al.* [49], show that Johnsongrass control could be as effective with imazethapyr as with haloxyfop, particularly when application timing coincides with a crucial period for the survival of the weed population.

Before the 90's, the use of glyphosate involved high costs and it was limited to selective applications or control of Johnsongrass in reduced areas. The massive adoption of glyphosate was linked to zero tillage and round-up ready soybeans during 90's and the problem of Johnsongrass in crop lands was reduced. However, continuous use of herbicides with the same mode of action is a key factor in the evolution of herbicide resistant weed populations. Vila-Aiub *et al.* [50] establish that Johnsongrass control failures at commercial field rates of glyphosate in Argentina have been a consequence of evolved heritable resistance to glyphosate. In consequence, this weed is again a threat to agriculture.

CYNODON DACTYLON: A COSMOPOLITE WEED

Cynodon dactylon (L.) Pers. (Bermudagrass) is a rizomatous perennial grass. This species is native from Africa but this weed is a cosmopolite plant due to the high adaptability to a wide range of environmental conditions. In South America, Bermudagrass distributed between 24°S to 42°S latitudes including a large part of Argentina and Uruguay. Several Cynodon dactylon sub-species and varieties had been identified because this species is variable in its ecological and morphological traits [19, 24].

Description and Life Cycle

Bermudagrass is a small plant (stems 10-50 cm tall) with mostly creeping and stoloniferous, branched, slightly flattened stems, flowering culms erects. The laminas 2-10 cm long by 0.2-0.4 cm wide. The stolons have short internodes, producing digitate inflorescences at the nodes. The inflorescence is 2.5 cm long with 3-8 spikes in a finger-like arrangement unifloras spikelets 0.2-0.3 cm long. This warm-season plant tolerates low temperatures and long periods of drought due to its deep and vigorous rhizomes and fibrous roots [24].

It is considered one of the world's worst weeds, and invades most subtropical and temperate agroecosystems in Argentina [51]. As for other C4 perennial grasses, the growth of Bermudagrass is modulated by temperature, and in Pampa region a winter rest period is observed [17]. This weed begins growth in the spring and continues during the summer months; initial spring growth uses its stored carbohydrate reserves of the rhizomes, mainly a starch [52].

Cynodon dactylon is out-crossers and self-sterile, but the seed production is generally low and seed viability varies with the biotype and environmental conditions, although seed production is important for invading new areas. In this sense, Bermudagrass reproduces mainly by rhizomes and stolons. Bud activity on nodes of stolons and rhizomes is regulated by apical dominance, which can be broken by sectioning the stolons or rhizomes through mechanically tillage [53].

In the winter, the plants remain alive and re-grow actively during the spring. The growth of underground parts starts earlier than that of the top. The main factors regulating sprout emergence dynamics seem to be temperature and distribution of weed structures in the soil. Buds do not sprout at 7°C and the base temperature is 8°C. At temperatures higher than 11°C until 33°C, the sprout of bud responds positively to the temperature [54]. In established plants, stolon elongation is intensive in spring. Under normal cropping conditions, more branches can be originated from rhizome than from equivalent stolon fragments [17]. Flowering culms appear in middle of the summer and these fruits until late autumn. According to Horowitz [55], most plants rhizome initiation preceded flowering, however no relationship was found between flowering or aerial development and rhizome development. Finally, the first frosts kill the aboveground organs [24].

Cynodon dactylon is a highly plastic species in response to growth factors as the incident irradiance [56]. Beltrano et al. [57] have found that a relatively high photosynthetic activity (high leaf area under high photon flux) of the plant produced enough assimilates to induce the postrate growth of the stolon grown either under low and high *far red/red* irradiation. This must occur in the filed under full sunlight. Low photon flux or reduced leaf area might diminish the amount of photoassimilates to a level where distribution would depend upon the *far red/red* ratio received by the stolon. When the plant had low assimilates content, sugar partitioning was mainly towards those stolons grown under low *far red/red* in detriment of the stolons irradiated with high *far red/red*. This could be an important factor in the growth responses of the weed under light competition.

On the other side, other researches about the growth of Bermudagrass green area in the field have shown that the spatial growth of Bermudagrass patch is characterized by the centrifugal extension of primary stolons from the original seed patch. Under full sunlight, it allocates biomass mainly to stolons and proportionally less to leaves and the spatial growth is described by concentric growth of a moving front constituted by numerous packed stolons of several categories (primary stolons highly branched) that engulf the area. The shading reduces the number and extension rate of stolons. Patches lost their elliptical shape because stolons are widely spaced. This patch growth pattern show few primary stolons that are unbranched. Thus, biomass production and partitioning regulates the spatial growth and morphological plasticity of *Cynodon dactylon* patches [16]. Consistent, Dong and de Kroon [58] have reported that stolon and rhizome branching intensities are reduced in response to lower light and lower nutrient levels. Stolons and their internodes elongate greatly under lower light level, but slightly shorten under lower nutrient levels. In addition, they found that the lengths of rhizomes and their internodes do not respond to nutrient availability. Thus, the morphology of stolons might be more responsive than the morphology of rhizomes to resource supply.

Water is another important factor affecting *Cynodon dactylon* growth. Bermudagrass patch growth and biomass production are greatly reduced by low soil water regimes. According to De Abelleyra et al. [59], reductions in final green area are of greater magnitude than reductions in biomass production. This could be an indicator that *Cynodon dactylon* area increment is more sensitive to water stress than biomass production, probably due to differences between leaf extension and biomass production response to soil moisture condition.

Anatomical changes certainly play a crucial role in combination with physiological attributes in response to stress factors. According to Hameed et al. [60], salt tolerant ecotypes increase exodermis and sclerenchyma in roots for preventing water loss through root surface, endodermis for preventing redial flow of water and nutrients, and increase cortex and pith parenchyma for better water storage. Furthermore, these ecotypes accumulate higher organic osmotic (total free amino acids, proline, total soluble proteins, and total soluble sugars) under saline conditions than non adapted salt ecotypes [61].

The *Cynodon dactylon* responses to growth factors not only have ecological implications in the adaptation along to several environments, but also ecotypes differences and morphologic changes are related to weed-crop interference. Ramakrishnan and Gupta [62] have found differential competitive interferences among Bermudagrass populations. Such a differential behavior of the weed has been related to the eco-physiological make-up of the ecotype populations. According to Ott [52], the genetic variability among the ecotypes and the different expressions of these types in different environments suggest that in weed-crop studies the conventional species cannot be treated as a homogeneous entity and make it necessary to seek local or regional information for integrated management of this weed on a permanent basis.

Weed-crop Interference and Control of Bermudagrass

In Argentina, yield losses were estimated about 44% in soybean, 35% in sunflower, and 39% in potato when crops competed all season with Bermudagrass [51]. On the other hand, the reduction of 72% of maize root dry weight was particularly severe. It is quite clear that the main effect of *C. dactylon* on maize was exerted via the roots [63].

The Bermudagrass-crop interference could be assigned to competitive and allelopathic effects. Vasilakoglou et al. [64] based on the results of the laboratory experiment indicate that inhibitory substances are present in extracts of Bermudagrass rhizomes and foliage. These substances have strong potential for resource-based competition on corn grown under field conditions and could potentially influence initial growth and yield of maize. Also, the resource-based interference during the first four weeks after crop planting is able to reduce significantly the yield of maize. Therefore, control of this perennial weed species should be done within this time to avoid their competition or possible release of their allelopathic substances.

Previous to the diffusion of zero-tillage, soil ploughing and disking were usually practised as Bermudagrass control method under conventional tillage systems, it produces fragmentation and soil burial of both overwintering stolons and rhizomes of *Cynodon dactylon*. In those agricultural systems where the soil is not disturbed by mechanical cultivations, the value of stolon fragments as vegetative propagules of the weed may increase relatively to the intact rhizomes. Under these conditions, the best strategy to slow the expansion of the weed would require the avoidance of stolon fragmentation and dispersal, and also to eliminate most of the stolon and rhizome buds by applying systemic herbicides [17]. In this sense, among the agrochemical products used to control *Cynodon dactylon*, glyphosate is commonly use to its control in fallows. High levels of Bermudagrass control are obtained ACCase inhibitors as fenoxaprop-P-ethyl, haloxyfop, propaquizafop, and quizalofop-P-ethyl during postemergence of crops [51]. However, the moment of the control based in the biologic traits of this weed might be an important factor for the success of the applications.

Guglielmini et al. [65] based upon analysis of *Cynodon dactylon* population dynamics have determined two critical periods for the control of this weed:

Early critical control period: It has been established when the weed shows initially a low growth and the sprout of buds is linked with a high energetic cost. The efficacy of the control techniques is high, but it decreases if the practice is postponed. In this period, the control percentage of Bermudagrass using herbicides was greater than 60%.

Late critical control period: This period has been placed towards the final of Bermudagrass growth cycle when the assimilate distribution changes from aboveground to belowground organs. The goal is to prevent the growth of new rhizomes the following spring. It is mainly supported by the control technique in the application of herbicides that stop early the end of the growth cycle. The results of efficacy obtained in this period might be variable compared to the early critical control period.

In all cases, the strategy of Bermudagrass management should be defined in terms of a program that combines control methods and a long-term follow-up. These aspects are discussed below.

LOLIUM PERENNE POPULATIONS,
A WEED OF CEREAL CROPS IN ARGENTINA

In Argentina, naturalized populations of *Lolium spp.* complex (ryegrass) grow in disturbed plant communities throughout the country and have been present in grasslands of the Pampa region [66]. *Lolium* species are native from Europe, temperate Asia and North Africa although most species have been widely distributed around the temperate areas of the world. Within the genus *Lolium*, two groups are clearly distinguished based on morphological and phenological data: one group, containing the two autogamous species *L. temulentum* and *L. persicum*; the second group, the out-breeding species, *L. perenne, L. rigidum*, and *L. multiflorum* [67].

The two most important ryegrass species are *L. multiflorum* and *L. perenne*, which widely grow as forage and cover crops as well as turf grasses. These species can become serious weeds when they escape from fields [68]. *L. multiflorum* is an annual plant, occasionally biannual, while *L. perenne* is a perennial plant [66].

During several decades, ryegrass was planted in many systems of cattle production from Buenos Aires province. Subsequently, the rise of agriculture led to the change in land use and many systems turned to grain production. Then, *Lolium spp.* was an important weed in this new context.

Description and Life Cycle

Perennial ryegrass has been found as weed in agroecosystems from the south of Buenos Aires province. Often, this species is associated to *L. multiflorum* as components of the weed community of the region. However, several differences are distinguished *L. perenne* from *L. multiflorum*: folded vs rolled leaf vernation, growth habit, requirement of vernalization and number of spikelets by spike [66, 69, 70]. Anyway, the production of hybrids between these species is frequent and some authors concluded that cross-breeding species should be a same species [71]. Although many aspects discussed here might be similar to *L. multiflorum*, this section is based on the biology of *L. perenne*.

Perennial ryegrass is a plant of up 80 cm tall. The leaf blades are 20-30 cm of length and 2-6 mm of width. Tillers show folded leaf vernation. Short rhizomes are formed from rooting of basal nodes of culms. Inflorescens are normally unbranched, mostly 10-20 cm long with 20-30 spikelets of up 10 florets, lemmas without awn. Sipekelet stand up in third on the glume. Sterile tillers surround flowering culms. The high plasticity of this species makes those theses parameters are affected according to the conditions of growth.

In the south of Buenos Aires province, perennial ryegrass seed generally germinates with the first autumn rains, but it requires previous release of the dormancy conditioned to thermal after-ripening time. Thus, a proportion of the seed population still remains ungerminated after the first rains. Growth rate of the seedlings is initially low, when the plant has four expanded leaves the tillering period starts, unexpanded leaves and apical meristems are strong sinks of assimilates and the rate growth increase. This species has an obligatory requirement for exposure to both vernalization and long day for its inflorescence initiation [72]. Flowering occurs staggered in spring while the production of tillers still continues during this stage.

Initially, spikelets begin the anthesis from the florets of the middle to extrems. Self-incompatibity provokes an obligate out-breeding reproductive habit. Towards the end of spring, with increments in temperatures, grains mature and begin the dehiscence; potential seed yield is near to 10,000 seeds plant[-1]. Seeds enter the soil from this source, *i.e.,* it is plants that escape control and produce seeds within the field. Seeds of *L. perenne* are low persistent and survival rate after one year is generally very low if the seeds are left near the soil surface for approximately one month before deeper incorporation [73]. In the summer, plants cease growth completely and most mature aerial tissues senesce, but below-ground organs are kept dormant. On the next season, perennial ryegrass re-grows vigorously and new seeds sprout with the autumn rains and the cycle continues.

Ryegrass-crop Interference

Ryegrass species are considered important weeds in different cropping systems from Argentina. *L. multiflorum* and *L. perenne* can be found affecting wheat, barley and oats, where chemical control is the main strategy used [13]. In wheat, yield losses were recorded about 40-50 % due to ryegrass-crop competition [74].

In an effort to describe the interference between wheat and ryegrass, Hashem et al. [75] found that wheat was the stronger competitor during vegetative stages, but ryegrass became the stronger competitor during the reproductive stages of development. In this sense, wheat leaves dominated at the top canopy during the vegetative stage, but ryegrass dominated at the top canopy during the reproductive stages.

Ryegrass produced significant reductions in shoot dry matter yield and grain yield of wheat. Acciaresi et al. [74] have found that these effects are distinguished according to: root or shoot competition. The root competition was higher than shoot competition, causing shoot dry matter yield losses by 49 % and a grain number ear[-1] reduction by 35% while shoot competition reduced the grain size by 11 %. These effects were different among wheat cultivars evaluated.

Gussin and Lynch [76] reported the allelopathic interference of perennial ryegrass residues. This weed can inhibit Bermudagrass germination and growth in soil amended with shoot residues [77]. Furthermore, Mattner and Parbery [78] found that rust may enhance ryegrass allelopathy against clover. Allelochemicals were isolated and identified from the aqueous leachates of decaying *L. multiflorum* residues. One of the allelochemicals was identified as benzenepropanoic acid; this inhibited the elongation of rice seedling roots [79]. Amini et al. [80] have communicated that the inhibitory effect of ryegrass on wheat root and shoot length was chemically directed through its root exudates and growth inhibition was greater on wheat roots than on wheat shoots. In addition, this allelopathic effect of ryegrass differed on root and shoot growth and among wheat cultivars.

Varieties with high weed-suppressing potential are likely to play an important role in natural weed control [81]. Cultivars that provide early competition are an important factor when a program of integrated weeds management is designed.

Glyphosate-resistant Perennial Ryegrass

In the south of Buenos Aires province a population of Perennial ryegrass was identified as glyphosate-resistant based on the poor control at commercial glyphosate doses. This problem population arose in a field with a history of twelve years under non-tillage agriculture, with weed control based on three applications of glyphosate per year at doses of 360 to 720 g ae ha^{-1} [82].

Glyphosate effects on physiologic process were studied and compared susceptible and glyphosate-resistant biotypes. The herbicide did not influence assimilates translocation, or release of root exudates in the resistant population. The translocation of assimilates and its distribution patterns were significantly affected by glyphosate within one day in the susceptible population. The treated susceptible plants showed 57% higher assimilates retention at the expanded leaves than their controls. The lower assimilates movement significantly affected the unexpanded leaves and the apical meristems. Moreover, the exudation released from roots was significantly decreased by glyphosate only in the susceptible plants [83].

In the glyphosate-resistant *L. perenne* biotype, stomatal conductance was the only parameter slightly affected only 5 days post-application of glyphosate. In susceptible Perennial ryegrass biotypes, accumulation of reduced carbohydrates occurred before a decrease in stomatal conductance and CO_2 assimilation. The gas exchange was reduced earlier than chlorophyll fluorescence and the amount of chlorophyll in susceptible plants. In this sense, the initial glyphosate effects on gas exchange could be a response to a feedback regulation of photosynthesis. As discussed above, the herbicide affects actively growing tissues regardless of the inhibition of photosynthesis, the demand of assimilates decreased and consequently induced an accumulation of carbohydrates in leaves [84, 85].

Currently, around 8,000 ha are infested with this weed in the mentioned region; the management is based on the control with haloxyfop-R-methyl and clethodim at doses recommended by the manufacturers [82]. The physiologic characterization of the glyphosate-resistant biotypes could be a contribution toward knowing the mechanism of glyphosate-resistance.

INTEGRATED PERENNIAL GRASS WEEDS MANAGEMENT

An integrated weed management program combines control methods to reduce competition with the crop. The methods may be preventive, cultural, mechanical, or chemical. The goal of an integrated program is to give reliable, effective weed control while minimizing environmental hazards.

Prevention

It is less expensive and time-consuming to keep Johnsongrass, Bermudagrass and Perennial ryegrass out of a field than to control these weeds once they are established. Mapping and monitoring of weed populations is necessary in this sense. To prevent an infestation, an only certified weed-free seeds must be sowed. Control Johnsongrass, Bermudagrass and Perennial ryegrass in non-crop and fencerows areas to reduce sources of weed seeds.

The equipment must be driven around, rather than through, isolated patches of weeds. To avoid spreading vegetative propagules, thoroughly clean equipment (especially combines) after working in infested fields. Moreover, infested fields must be harvested last so that seeds will not be transported into other areas.

Cultural Control

The following cultural practices help crops compete with perennial grass weeds:

- Soil-test recommendations for fertilizer must be followed.
- Crop sowing must be performed as soon as soil temperatures are optimal.
- When possible, plant arrangement and density must be modified to confer a higher competitive ability to the crop.
- High-yielding genotypes adapted to local climate, soil, and field conditions must be sowed.
- Regularly field scouting for weeds must be performed and then chemical control must be performed when necessary.
- Crops that provide early competition (such as small grains) must be included in the rotation.

Mechanical Control

Mechanical control methods include hand-pulling, hoeing, mowing, plowing, disking, and cultivating. While hand-pulling and hoeing are useful for controlling individual plants or small weed patches, these methods are too time-consuming and laborious to be economical on a large scale. Mowing or harvesting prevent weed seed production in small grains, pastures, and non-crop areas, but it is not suitable for corn and soybean fields.

Fall plowing, where appropriate, will expose Johnsongrass and Bermudagrass rhizomes to killing temperatures. If fall plowing is not possible, spring plowing as soon as the soil is workable must be performed.

Johnsongrass and Bermudagrass rhizomes must be disking, to make them more susceptible to herbicides. Disk to a 15 to 20 cm depth several times before planting, and use an herbicide program that is effective on both grasses. Disking alone can spread rhizome fragments, so be sure to take appropriate follow-up measures. Moreover, Perennial ryegrass seed succumb early after storage at 15 cm depth in soil.

Cultivation reduces carbohydrate reserves in Johnsongrass and Bermudagrass, making them less competitive. Cultivating controls weeds between crop rows, but does not kill weeds near crop plants. Cultivating two or three times during the first six weeks after planting will keep weeds in check between rows until the canopy is established. Tillage equipment must always be cleaned after working in one area and before moving into another.

Chemical Control

Herbicides can be a useful tool in a weed control program when combined with preventive, cultural, and mechanical methods. To ensure that the herbicides used are as effective, safe, and economical as possible, always:

- An appropriate herbicide for specific weed species and crop must be selected. Stage of crop and weed growth, soil moisture, and temperature can affect herbicide selection.
- The product label directions must be carefully considered.
- Specific herbicides must be applied at the proper time.
- The recommended rate to avoid injury, residues, or poor control must be applied.
- Use multiple herbicide sites-of-action with overlapping weed spectrums in rotation.
- Application equipment must be calibrated several times during the season to ensure that the correct amount of herbicide is applied.
- Fields must be regularly scouted and the types and locations of weeds present must be recorded. These records can help to design an integrated control program for perennial grass weeds.

CONCLUSION

In summary, several methods could be suggested for the implementation of an integrated weed management program; however, these recommendations should be based on biological traits of weeds. The high plasticity of perennial weeds to adapt to different environments and control practices requires promoting the competitive ability of crops rather than eradication of weeds.

REFERENCES

[1] J. Dekker, Weed Diversity and Weed Management, *Weed Science.* 45 (1997) 357-363.
[2] C.M. Ghersa, M.L. Roush, S.R. Radosevich, S.M. Cordray, Coevolution of Agroecosystems and Weed Management, *BioScience.* 44 (1994) 85-94.
[3] R.. Froud-Williams, Changes in weed flora with different tillage and agronomic management systems, in: M.A. Altieri, M. Liebman (Eds.), Weed Management in Agroecosystems: Eco- Logical Approaches., CRC Press, Boca Raton, 1988: pp. 213-236.
[4] R. Cousens, M. Mortimer, Dynamics of weed populations, Cambridge University Press, Cambridge, 1995.
[5] W. Pengue, Transgenic crops in Argentina: the ecological and social debt., *Bulletin of Science, Technology and Society.* 25 (2005) 314-322.
[6] E. Trigo, E. Cap, V. Malach, F. Villarreal, The Case of Zero-Tillage Technology in Argentina, 2020th ed., IFPRI, Washington, 2009.

[7]   A.S. Grandy, G.P. Robertson, K.D. Thelen, Do Productivity and Environmental Trade-offs Justify Periodically Cultivating No-till Cropping Systems?, *Agronomy Journal*. 98 (2006) 1377.

[8]   E. Bonari, M. Mazzonzini, A. Peruzzi, M. Ginanni, Soil erosion and nitrogen loss as affected by different tillage systems used for durum wheat cultivated in a hilly clayey soil, in: F. Tebrügge, A. Böhrnsen (Eds.), Experiences with the Applicability of No-tillage Crop Production in the West-European Countries, Proc. EC-Workshop II, Giessen, 1995: pp. 81-91.

[9]   W.G. Johnson, V.M. Davis, G.R. Kruger, S.C. Weller, Influence of glyphosate-resistant cropping systems on weed species shifts and glyphosate-resistant weed populations, *European Journal of Agronomy*. 31 (2009) 162-172.

[10]   E. Puricelli, D. Tuesca, Weed density and diversity under glyphosate-resistant crop sequences, *Crop Protection*. 24 (2005) 533-542.

[11]   E. de la Fuente, S. Suárez, C. Ghersa, R. León, Soybean Weed Communities/: Relationships with Cultural History and Crop Yield, *Agronomy Journal*. (1999) 234-241.

[12]   E.B. de la Fuente, S.A. Suárez, C.M. Ghersa, Soybean weed community composition and richness between 1995 and 2003 in the Rolling Pampas (Argentina), Agriculture, Ecosystems andamp; *Environment*. 115 (2006) 229-236.

[13]   C. Istilart, M. Yanniccari, Análisis de la evolución de malezas en cereales de invierno durante 27 años en la zona sur de la pampa húmeda argentina., Revista Técnica Especial: Malezas Problema (Aapresid). (2011) 47-50.

[14]   M.M. Vila-Aiub, M.C. Balbi, P.E. Gundel, C.M. Ghersa, S.B. Powles, Evolution of Glyphosate-Resistant Johnsongrass (Sorghum halepense) in Glyphosate-Resistant Soybean, *Weed Science*. 55 (2007) 566-571.

[15]   I. Olea, Situación de la resistencia y control de malezas resistentes a herbicidas en la Argentina, in: Coloquio Sobre Resistencia De Malas Hierbas a Herbicidas, 2011: p. 11.

[16]   A.C. Guglielmini, E.H. Satorre, Shading effects on spatial growth and biomass partitioning of Cynodon dactylon, *Weed Research*. 4453 (2002) 123-134.

[17]   O.N. Fernandez, Establishment of Cynodon dactylon from stolon and rhizome fragments, *Weed Research*. (2003) 130-138.

[18]   A.E. Abdullahi, Cynodon dactylon control with tillage and glyphosate, *Crop Protection*. 21 (2002) 1093-1100.

[19]   H.G. Baker, The Evolution of weeds, *Annual Review of Ecology and Systematics*. 5 (1974) 1-24.

[20]   Hakansson, Weeds and weed management on arable land. *An ecological approach*. CABI Publishing, Cambridge, 2003.

[21]   J. Lindquist, Mechanisms of crop loss due to weed competition., in: R.K.D. Peterson, L.G. Higley (Eds.), Biotic Stress and Yield Loss, CRC Press, Boca Raton, 2001: pp. 233-253.

[22]   D.E. Goldberg, K. Landa, Competitive effect and response: hierarchies and correlated traits in the early stages of competition, *Journal of Ecology*. 79 (1990) 1013-1030.

[23]   J.D. Snowden, The Cultivated Races of Sorghum, Allard and Son, London, 1936.

[24]   A. Marzocca, O. Marsico, O. del Puerto, Manual de Malezas, Hemisferio Sur, Buenos Aires, 1976.

[25]   N. Monaghan, The biology of Johnson grass (Sorghum halepense), *Weed Research*. 19 (1979) 261-267.

[26]   McWhorther, Morphology and development of Johnsongrass plants from seeds and rhizomes, *Weeds*. 9 (1961) 558-562.

[27]   C.M. Ghersa, E.H. Satorre, M.L. van Esso, Seasonal patterns of Johnsongrass seed production in different agricultural systems, *Israel Journal of Botany*. 34 (1985) 24-31.

[28]   C. McWhorther, Growth and Development of Johnsongrass Ecotypes, *Weed Science*. 19 (1971) 141-147.

[29]   M. Horowitz, Seasonal Development of Established Johnsongrass, *Weed Science*. 20 (1972) 392-395.

[30]   I.M. Wedderstoon, W. Bustg, Growth and develop- ment of three johnsongrass selections, *Weed Science*. 22 (1974) 319-322.

[31]   M. Horowitz, Early Development of Johnsongrass, *Weed Science*. 20 (1972) 271-273.

[32]   A. Mitidieri, Importance, biology and basic control of Sorghum halepense (L.) Pers., in: Ecology and Control of Perennial Weeds in Latin America., FAO, Rome, 1986: pp. 1-35.

[33]   M. Czarnota, R.N. Paul, F.C. Dayan, C.I. Nimbal, L.A. Weston, Mode of action, localization of production, chemical nature, and activity of sorgoleone: a potent psII inhibitor in Sorghum spp. root exudates, (n.d.).

[34]   C.H. Nimbal, J.F. Pedersen, C. Yerkes, L.A. Weston, C.S. Weller, Phytotoxicity and distribution of sorgoleone in grain sorghum germplasm, *Journal of Agricultural and Food Chemistry*. 44 (n.d.) 1343-1347.

[35]   H.A. Acciaresi, C.A. Asenjo, Efecto alelopático de Sorghum halepense (L.) Pers. sobre el crecimiento de la plántula y la biomasa aérea y radical de Triticum aestivum (L.) , *Ecología Austral*. 13 (2003) 49-61.

[36]   C.M. Ghersa, M.A. Martínez-Ghersa, E.H. Satorre, M.L. Van Esso, G. Chichotky, Seed dispersal, distribution and recruitment of seedlings of Sorghum halepense (L.) Pers., *Weed Research*. 33 (1993) 79-88.

[37]   C.M. Ghersa, M.A. Martinez-Ghersa, A Field Method for Predicting Yield Losses in Maize Caused by Johnsongrass (Sorghum halepense), *Weed Technology*. 5 (1991) 279-285.

[38]   D.T. Patterson, Effects of environmental stress on weed/crop interactions, *Weed Science*. 43 (1995) 483-490.

[39]   E.S. Leguizamón, M.E. Yanniccari, J.J. Guiamet, H.A. Acciaresi, Growth , gas exchange and competitive ability of water availability, *Canadian Journal of Plant Science*. 91 (2011) 1011-1025.

[40]   M.S. Heschel, S.E. Sultan, S. Glover, D. Sloan, Population differentiation and plastic responses to drought stress in the generalist annual, Polygonum persicaria., *International Journal of Plant Science*. 165 (2004) 817-824.

[41]   J.P. Sexton, J.K. McKay, A. Sala, Plasticity and genetic diversity may allow saltcedar to invade cold climates in North America, *Ecological Applications*. 12 (2002) 1652-1660.

[42]   H.A. Acciaresi, J.J. Guiamet, Below- and above-ground growth and biomass allocation in maize and Sorghum halepense in response to soil water competition, *Weed Research*. 50 (2010) 481-492.

[43] C. McWhorther, History, biology and control of Johnsongrass, *Weed Science.* 4 (1989) 85–121.

[44] C.G. McWhorter, W.L. Barrentine., Weed control in soybeans with mefluidide applied postemergence, *Weed Science.* 27 (1979) 42-47.

[45] W.L. Barrentine., C.G. McWhorter, Johnsongrass (Sorghum halepense) Control with Herbicides in Oil Diluents, *Weed Science.* 36 (1988) 102-110.

[46] P.R. Vidrine, Johnsongrass (Sorghum halepense) control in soybeans (Glycine max) with postemergence herbicides, *Weed Technology.* 3 (1989) 455–458.

[47] P. Leiva, N. Ianonne, Soja: el problema de las malezas y su control: 2ª parte., Carpeta De Producción Vegetal. EEA.INTA Pergamino. Información No. 119. (1996).

[48] M.E. Yanniccari, H.A. Acciaresi, Response of Johnsongrass Biotypes from Humid and Subhumid Regions to Nicosulfuron, *Crop Management.* (2012) doi:10.1094/CM-2012-0412-01-RS.

[49] H.J. Tassara, J. Santoro, M.C. de Seiler, E. Bojanich, C. Rubione, R. Pavon, et al., Johnsongrass (Sorghum halepense) control with imazethapyr and haloxyfop in conventional and vertical-tilled soybean (Glycine max), *Weed Science.* 44 (1996) 345-349.

[50] M.M. Vila-aiub, R.A. Vidal, M.C. Balbi, P.E. Gundel, F. Trucco, C.M. Ghersa, Glyphosate-resistant weeds of South American cropping systems : an overview, *Pest Management Science.* 371 (2008) 366-371.

[51] F. Bedmar, Bermudagrass (Cynodon dactylon) Control in Sunflower (Helianthus annuus), Soybean (Glycine max), and Potato (Solanum tuberosum) with Postemergence Graminicides, *Weed T.* 11 (1997) 683-688.

[52] P.M. Ott, Biology and ecology of Cynodon dactylon L. Pers, *Ecology and Control of Perennial Weeds in Latin America.* 74 (1983) 36-47.

[53] I. Moreira, Propagaçao por semente do Cynodon dactylon (L.) Pers, Anais Do Instituto Superior De Agronomía. 35 (1975) 95–112.

[54] E.H. Satorre, F.A. Rizzo, S.P. Arias, C.D. Cereaiicultura, P. Vegetal, V. Esso, The effect of temperature on sprouting and early establishment of Cynodon dactylon, *Weed Research.* 36 (1996) 431-441.

[55] M. Horowitz, Development of Cynodon dactylon (L.) Pers., *Weed Research.* 12 (1972) 207-220.

[56] M. Dong, M.G. Pierdominici, Morphology and growth of stolons and rhizomes in three clonal grasses, as affected by different light supply, *Vegetation.* 116 (1995) 25-32.

[57] J. Beltrano, J. Willemoës, E.R. Montaldi, R. Barreiro, Photoassimilate partitioning modulated by phytochrome in Bermuda grass (Cynodon dactylon (L) Pers.), *Plant Science.* 73 (1991) 19-22.

[58] M. Dong, H. de Kroon, Plasticity in Morphology and Biomass Allocation in Cynodon dactylon, a Grass Species Forming Stolons and Rhizomes, *Oikos.* 70 (1994) 99-106.

[59] D. De Abelleyra, A.M. Verdú, B.C. Kruk, E.H. Satorre, Soil water availability affects green area and biomass growth of Cynodon dactylon, *Weed Research.* 48 (2008) 248-256.

[60] M. Haasan, M. Ashraf, N. Naz, F. Al-Qurainy, Anatomical adaptations of Cynodon dactylon (L.) PERS., from the salt range Pakistan, to salinity stress. I. root and stem anatomy, *Pakistan Journal of Botany.* 42 (2010) 279-289.

[61] M. Haasan, M. Ashraf, Physiological and biochemical adaptations of Cynodon dactylon (L.) Pers., from the Salt Range (Pakistan) to salinity stress, *Flora*. 203 (2008) 683-694.

[62] P.S. Ramakrishnan, U. Gupta, Ecotypic Differences in Cynodon dactylon (L.) Pers. Related to Weed-Crop Interference, *Journal of Applied Ecology*. 9 (1972) 333-339.

[63] D. Abdul Shukor Juraimi, S. H Drennan, N. Anuar, Competitive Effect of Cynodon dactylon (L.) Pers. on Four Crop Species, Soybean [Glycine max (L.) Merr.], Maize (Zea mays), Spring Wheat (Triticum aestivum) and Faba Bean [Vicia faba (L.)]., *Asian Journal of Plant Sciences*. 4 (2005) 90-94.

[64] I. Vasilakoglou, K. Dhima, I. Eleftherohorinos, Allelopathic Potential of Bermudagrass and Johnsongrass and Their Interference with Cotton and Corn, *Agronomy Journal*. 97 (2005) 303–313.

[65] A. Guglielmini, D. Batlla, R. Benech Arnold, Bases para el control y manejo de malezas., in: H. Satorre, R.L. Vence, G. Slafer, E. de la Fuente, D. Miralles, M.E. Otegui, et al. (Eds.), Producción De Granos. Bases Funcionales De Para Su Manejo, Editorial Facultad de Agronomía, 2003: p. 784.

[66] A. Cabrera, E. Zardini, Manual de la flora de los alrededores de Buenos Aires, *ACME*, 1978.

[67] B.P. Loos, Morphological variation in Lolium (Poaceae) as a measure of species relationships, *Plant Systematics and Evolutions*. 188 (1993) 87-99.

[68] K. Polok, Molecular evolution of the genus Lolium L., SQL, Olsztyn, 2007.

[69] G. Jung, A. Van Wijk, W. Hunt, C. WATSON, Ryegrasses, in: L. Moser, D. Buxton, M. Casler (Eds.), Cool-season Forage Grasses, American Society of Agronomy, Crop Science Society of America, Soil Science Society of America, Madison, 1996: p. 841.

[70] L.A. Inda Aramendía, El género Lolium. Claves dicotómicas., Revista Real Academia De Ciencias. *Zaragoza*. 60 (2005) 143-155.

[71] Z. Bulinska-Radomska, R.N. Lester, Relationships between five Lolium species (Poaceae), *Plant Systematics and Evolution*. 148 (1985) 169-176.

[72] C.P. MacMillan, C.A. Blundell, R.W. King, Flowering of the Grass Lolium perenne. Effects of Vernalization and Long Days on Gibberellin Biosynthesis and Signaling, *Plant Physiology*. 138 (2005) 1794-1806.

[73] P.K. Jensen, Longevity of seeds of Poa pratensis and Lolium perenne as affected by simulated soil tillage practices and its implications for contamination of herbage seed crops, *Grass and Forage Science*. 65 (2010) 85-91.

[74] H.A. Acciaresi, H.O. Chidichimo, S.J. Sarandón, Shoot and Root Competition in a Lolium multiflorum-Wheat Association, *Biological Agriculture and Horticulture*. 21 (2003) 15-33.

[75] A. Hashem, S.R. Radosevich, M.L. Roush, Effect of Proximity Factors on Competition between Winter Wheat (Triticum aestivum) andItalian Ryegrass (Lolium multiflorum), *Weed Science*. 46 (1998) 181-190.

[76] E.J. Gussin, J.M. Lynch, Microbial Fermentation of Grass Residues to Organic Acids as a Factor in the Establishment of New Grass Swards, *New Phytologist*. 89 (1981) 449-457.

[77] L.B. McCarty, R.K. McCauley, H. Liu, F.W. Totten, J.E. Toler, Perennial Ryegrass Allelopathic Potential on Bermudagrass Germination and Seedling Growth , *HortScience* . 45 (2010) 1872-1875.

[78] S.W. Mattner, D.G. Parbery, Crown rust affects plant performance and interference ability of Italian ryegrass in the post-epidemic generation, *Grass and Forage Science*. 62 (2007) 437-444.

[79] L. Guoxi, Z. Sen, L. Houjin, Y. Zhongyi, X. Guorong, Y. Jiangang, et al., Allelopathic effects of decaying Italian ryegrass (Lolium mutiflorum Lam.) residues on rice, *Allelopathy Journal*. 22 (2008) 15-24.

[80] R. Amini, M. An, J. Pratley, S. Azimi, Allelopathic assessment of annual ryegrass (Lolium rigidum): Bioassays, *Allelopathy Journal*. 24 (2009) 67-76.

[81] M. Inderjit, J. Streibig, Wheat (Triticum aestivum) Interference with Seedling Growth of Perennial Ryegrass (Lolium perenne): Influence of Density and Age, *Weed Technology*. 15 (2001) 807-812.

[82] M. Yanniccari, C. Istilart, D.O. Giménez, A.M. Castro, Glyphosate resistance in perennial ryegrass (Lolium perenne L.) from Argentina, *Crop Protection*. 32 (2012) 12-16.

[83] M.E. Yanniccari, C. Istilart, D.O. Giménez, A.M. Castro, Effects of glyphosate on the movement of assimilates of two Lolium perenne L. populations with differential herbicide sensitivity, *Environmental and Experimental Botany*. 82 (2012) 14-19.

[84] M.E. Yanniccari, C. Istilart, D.O. Giménez, H.A. Acciaresi, A.M. Castro, Efecto del glifosato sobre el crecimiento y acumulación de azúcares libres en dos biotipos de Lolium perenne de distinta sensibilidad al herbicida, *Planta Daninha*. 30 (2012) 155-164.

[85] M.E. Yanniccari, E. Tambussi, C. Istilart, A.M. Castro, Glyphosate effects on gas exchange and chlorophyll fluorescence responses of two Lolium perenne L. biotypes with differential herbicide sensitivity, *Plant Physiology and Biochemistry*. 57 (2012) 210-217.

In: Agricultural Research Updates. Volume 5
Editors: P. Gorawala and S. Mandhatri

ISBN: 978-1-62618-723-8
© 2013 Nova Science Publishers, Inc.

Chapter 3

Anti-inflammatory Effect of the Waste Components from Soybean (*Glycine max* L.) Oil Based on DNA Polymerase λ Inhibition

Yoshiyuki Mizushina[1,2,], Yoshihiro Takahashi[3],*
Isoko Kuriyama[1] and Hiromi Yoshida[1]

[1]Laboratory of Food and Nutritional Sciences,
Faculty of Nutrition, Kobe Gakuin University,
Nishi-ku, Kobe, Hyogo, Japan
[2]Cooperative Research Center of Life Sciences,
Kobe Gakuin University, Chuo-ku, Kobe, Hyogo, Japan
[3]Laboratory of Research and Development,
Oguraya Yanagimoto Co., Ltd., Higashi-Nada-ku,
Kobe, Hyogo, Japan

Abstract

During screening for selective DNA polymerase (pol) inhibitors, we purified compounds 1–3 from the waste extracts of soybean (*Glycine max* L.) oil (i.e., the gum fraction). These isolates were identified by spectroscopic analyses as glucosyl compounds, an acylated steryl glycoside [β-sitosteryl (6'-*O*-linoleoyl)-glucoside, compound 1], a steroidal glycoside (eleutheroside A, compound 2), and a cerebroside (glucosyl ceramide, AS-1-4, compound 3). Compound 1 exhibited a marked ability to inhibit the activities *in vitro* of mammalian Y-family pols (pols η, ι, and κ), which are repair-related pols, and observed the 50% inhibitory concentration (IC_{50}) to be 10.2–19.7 μM. Compounds 2 and 3 selectively inhibited the activity of eukaryotic pol λ, which is a

* Corresponding author: Yoshiyuki Mizushina, PhD., Laboratory of Food and Nutritional Sciences, Faculty of Nutrition, Kobe Gakuin University, 518 Arise, Ikawadani-cho, Nishi-ku, Kobe, Hyogo 651-2180, Japan; Tel.: +81-78-974-1551 (ext. 3232); fax: +81-78-974-5689. E-mail address: mizushin@nutr.kobegakuin.ac.jp (Y. Mizushina).

repair/recombination-related pol, with IC_{50} values of 9.1 and 12.2 μM, respectively. However, these three compounds did not influence the activities of the other mammalian pol species tested. In addition, these compounds had no effect on prokaryotic pols or other DNA metabolic enzymes, such as human DNA topoisomerases I and II, T7 RNA polymerase, T4 polynucleotide kinase, and bovine deoxyribonuclease I. Compounds 1–3 also exhibited no effects on the proliferation of several cultured human cancer cell lines. Compounds 2 and 3 suppressed the inflammation of mouse ear edema induced by TPA (12-*O*-tetradecanoylphorbol-13-acetate); therefore, the tendency for pol λ inhibition *in vitro* by these compounds showed a positive correlation with anti-inflammation *in vivo*. These results suggested that these glucosyl compounds from soybean waste extracts might be particularly useful for their anti-inflammatory properties.

Keywords: Soybean, glucosyl compounds, DNA polymerase, enzyme inhibitor, anti-inflammation

ABBREVIATIONS

BER, base excision repair
DMSO, dimethyl sulfoxide
dNTP, 2'-deoxynucleoside 5'-triphosphate
dsDNA, double-stranded DNA
dTTP, 2'-deoxythymidine 5'-triphosphate
IC_{50}, 50% inhibitory concentration
K_i, inhibition constant
K_m, Michaelis constant
LD_{50}, 50% lethal dose
MTT, 3-(4,5-dimethyl-2-thiazolyl)-2,5-diphenyl-2H tetrazolium bromide
pol, DNA-dependent DNA polymerase (EC 2.7.7.7)
TLS, translesion synthesis
Tm, melting temperature
TPA, 12-*O*-tetradecanoylphorbol-13-acetate
V_{max}, maximum velocity
XPV, Xeroderma pigmentosum variant

1. INTRODUCTION

DNA polymerase (DNA-dependent DNA polymerase [pol], EC 2.7.7.7) catalyzes the addition of deoxyribonucleotides to the 3'-hydroxyl terminus of primed double-stranded DNA (dsDNA) molecules [1]. As pols play important maintenance roles in key eukaryotic systems, such as DNA replication, DNA recombination, and DNA repair [2], pol inhibitors can be employed as anticancer chemotherapy agents because they inhibit cell proliferation. Based on pol inhibitors' strategic effects, we established an assay for pol inhibitors [3], and have been screening for mammalian pol inhibitors from natural products including food materials and components for over 15 years.

The human genome encodes at least 14 pols that carry out cellular DNA synthesis [4, 5]. Eukaryotic cells contain 3 replicative pols (α, δ, and ε), 1 mitochondrial pol (γ), at least 10 non-replicative pols (β, ζ, η, θ, ι, κ, λ, μ, ν, and REV1) [5, 6]. Pols have a highly conserved structure, with their overall catalytic subunits showing little variance among species; conserved enzyme structures are usually preserved over time because they perform important cellular functions that confer evolutionary advantages. On the basis of sequence homology, eukaryotic pols can be divided into four main families, termed A, B, X and Y [7]. Family A includes mitochondrial pol γ as well as pols θ and ν; family B includes the three replicative pols α, δ and ε; and pol ζ; family X consists of pols β, λ and μ; and last, family Y includes pols η, ι, and κ in addition to REV1.

During our pol inhibitor studies, we have discovered more than 100 inhibitors of mammalian pols [8, 9], and have found that pol λ-selective inhibitors, such as curcumin derivatives [10–12], display 12-O-tetradecanoylphorbol-13-acetate (TPA)-induced anti-inflammatory activity [13–15]. These results led to the consideration that selective inhibitors of mammalian pol species could be utilized as chemotherapy agents for not only anticancer effects but also anti-inflammation.

In this study, we screened pol inhibitors from the waste extract of soybean (*Glycine max* L.) oil—that is, the screw-pressed extracted components after soybean oil production—to develop bioactive foods and cosmetics from this industrial waste. We isolated three glucosyl compounds, as selective inhibitors of eukaryotic pol species, and investigated their anticancer and anti-inflammation bioactivities through evaluation of their pol inhibition activities.

2. PRODUCTION OF SOYBEAN OIL

The manufacturing process of soybean oil production is shown in the upper portion of Figure 1. First, the cleaned and dried soybeans are adjusted for moisture content and heated to coagulate soy proteins and facilitate oil extraction. The prepared soybeans are cut into flakes, screw-pressed, placed into percolation extractors, and combined with a solvent. The *n*-hexane/soybean oil mix is then separated from the flakes and transferred to evaporators where the oil and *n*-hexane are separated. The evaporated *n*-hexane is recovered and reused in future extraction processes, while the *n*-hexane free crude soybean oil is subjected to further refining. The crude soybean oil contains many impurities that need to be removed. Oil insoluble materials are removed through filtration, while oil soluble materials are removed through degumming, neutralizing, and bleaching. A stripping and/or deodorizing step completes the refining process [16].

We focused on the waste materials produced by the crude oil refining stage of the soybean oil production process, screened them for pol inhibitory activity, and found that the waste gum fraction produced by degumming inhibited the activity of eukaryotic pol species. In degumming, crude soybean oil and warmed-water are mixed and the resulting pellet of gum and solvent separated by centrifugation. In general, it is known that this gum pellet fraction contains phosphatides and gummy solids.

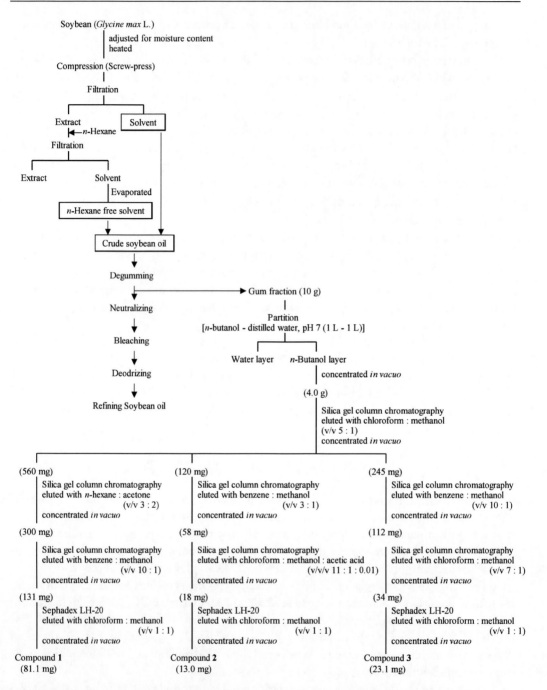

Figure 1. Schemes for soybean oil production and for purification of compounds 1–3, selective inhibitors of eukaryotic pol species, from the waste gum fraction produced in the degumming process during soybean oil refining.

3. ISOLATION OF MAMMALIAN POL INHIBITORS FROM THE WASTE GUM FRACTION PRODUCED BY DEGUMMING OF SOYBEAN OIL PRODUCTION

3.1. Purification of Pol Inhibitors

As the waste gum fraction from crude soybean oil possessed inhibitory activity against eukaryotic pols, this fraction (10 g) was partitioned between *n*-butanol and distilled water (1 L each), adjusted to pH 7, and the organic layer removed and evaporated. The fraction was first subjected to silica gel column chromatography eluted with chloroform/methanol (5/1, v/v), resulting in three active fractions, A to C. Fraction A (560 mg) was purified by elution through a second silica gel column eluted with *n*-hexane/acetone (3/2, v/v) and then a third silica gel column eluted with benzene/methanol (10/1, v/v). Fraction B (120 mg) was further purified by elution through a second silica gel column with benzene/methanol (3/1, v/v), followed by an additional silica gel column eluted with chloroform/methanol/acetic acid (11/1/0.01, v/v/v). Fraction C (245 mg) was further purified by elution through a silica gel column with benzene/methanol (10/1, v/v), then another silica gel column eluted with chloroform/methanol (7/1, v/v). Each chromatographed fraction was finally purified by elution through a Sephadex LH-20 (GE Healthcare Life Sciences, Ltd., Uppsala, SE) column eluted with chloroform/methanol (1/1, v/v). Finally, after solvent evaporation, white powders of compounds 1–3 were obtained (81.1, 13.0, and 23.1 mg, respectively) as pol inhibitors (Figure 1).

3.2. Structure Determination of Compounds 1–3

The chemical structures of the purified pol inhibitors, compounds 1–3, were analyzed by ^{1}H-NMR, ^{13}C-NMR, COSY, HMQC, and HMBC. ^{1}H and ^{13}C-NMR were recorded on a Bruker DRX400. Chemical shifts were reported in δ, parts per million (ppm), relative to tetramethylsilane as an internal standard. Mass spectra were obtained on API QStar Pulsar I spectrometer.

^{1}H and ^{13}C NMR spectral data confirmed the presence of a standard β-sitosteryl (6'-*O*-acyl)-glucoside in compound 1. ESIMS/MS analysis of the ion peak of compound 1 ([M+Na]$^+$) exhibited fragment ions corresponding to steryl glucoside, m/z 465 ([M-$C_{29}H_{49}$+Na]$^+$), and β-sitosterol, m/z 397 ([M-($C_6H_{10}O_6$-CO-$C_{17}H_{31}$)]$^+$). Thus, compound 1 was identified as an acylated steryl glycoside, β-sitosteryl (6'-*O*-linoleoyl)-glucoside (Figure 2A). NMR data matched those in ^{1}H and ^{13}C-NMR and MS data were consistent with reported values [17–19].

The molecular formulas of compounds 2 and 3 were determined to be $C_{35}H_{60}O_6$ and $C_{40}H_{75}NO_9$ by high resolution mass spectra. From the ^{1}H and ^{13}C NMR spectral data, compounds 2 and 3 were identified as eleutheroside A (Figure 2B) and a cerebroside, AS-1-4 (Figure 2C), respectively. These spectroscopic data were consistent with those previously reported [20–22].

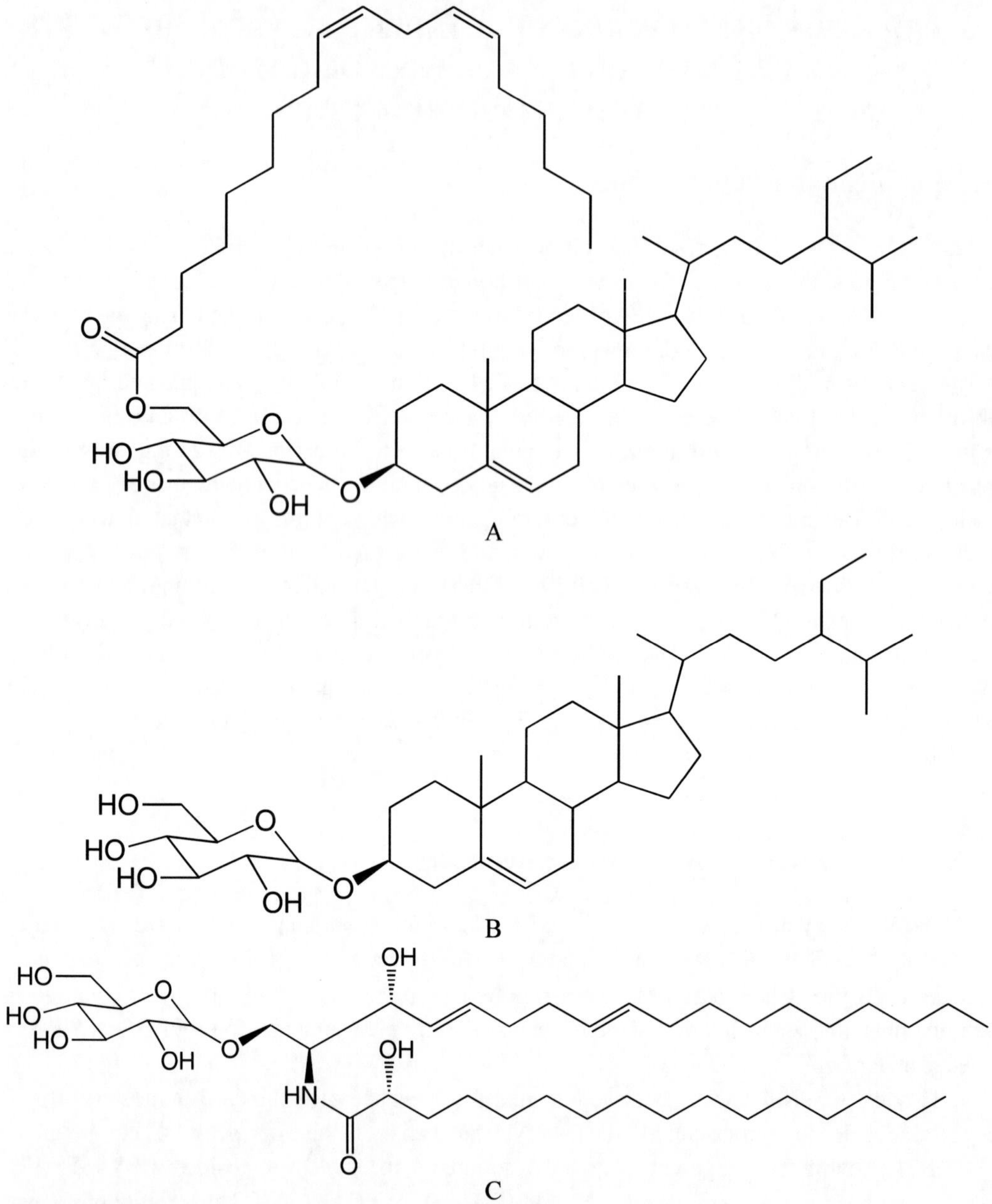

Figure 2. Structures of compounds 1–3. A, compound 1, an acylated steryl glycoside [β-sitosteryl (6'-*O*-linoleoyl)-glucoside; B, compound 2, a steroidal glucoside, eleutheroside A; and C, compound 3, cerebroside, AS-1-4.

As shown in Figure 2, compounds 1–3 are an acylated steryl glucoside, a steroidal glucoside, and a cerebroside (glucosyl ceramide), respectively, all having one glucose molecule in their structure.

4. EFFECTS OF COMPOUNDS 1–3 ON POL ACTIVITY

4.1. Preparation of Pols

Pol α was purified from calf thymus by immunoaffinity column chromatography, as described by Tamai et al. [23]. Recombinant rat pol β was purified from *E. coli* JMpβ5, as described by Date et al. [24]. The human pol γ catalytic gene was cloned into pFastBac. Histidine-tagged enzyme was expressed using the BAC-TO-BAC HT Baculovirus Expression System according to the supplier's instructions (Life Technologies, Inc., Gaithersburg, MD, USA) and purified using ProBoundresin (Invitrogen Japan K K, Tokyo, Japan) [25]. Human pols δ and ε were purified by nuclear fractionation of human peripheral blood cancer cells (Molt-4) using the second subunit of pol δ and ε-conjugated affinity column chromatography, respectively [26]. A truncated form of human pol η (residues 1–511) tagged with His$_6$ at its C-terminus was expressed in *E. coli* cells and purified as described by Kusumoto et al. [27]. A recombinant mouse pol ι tagged with His$_6$ at its C-terminus was expressed and purified by Ni-NTA column chromatography, as described elsewhere [Masutani et al., in preparation]. A truncated form of pol κ (residues 1–560) with His$_6$-tags attached at the C-terminus was overproduced in *E. coli* and purified, as described by Ohashi et al. [28]. Recombinant human His-pol λ was overexpressed and purified according to a method described by Shimazaki et al. [29]. Fish pol δ was purified from the testis of cherry salmon (*Oncorhynchus masou*) [30]. Fruit fly pols α, δ, and ε were purified from early embryos of *Drosophila melanogaster*, as described by Aoyagi et al. [31, 32]. Pol α from a higher plant, cauliflower inflorescence, was purified according to the methods outlined by Sakaguchi et al. [33]. Recombinant rice (*Oryza sativa* L. cv. Nipponbare) pol λ tagged with His$_6$ at the C-terminal was expressed in *E. coli* and purified from the cells, as described by Uchiyama et al. [34]. The Klenow fragment of pol I from *E. coli* was purchased from Worthington Biochemical Corp. (Lakewood, NJ, USA). T4 pol and *Taq* pol were purchased from Takara Bio, Inc. (Shiga, Japan).

4.2. Pol Assays

The reaction mixtures for calf pol α, rat pol β, plant pol α, and prokaryotic pols have been previously described [35, 36]. Those for human pol γ and for human pols δ and ε were as described by Umeda et al. [25] and Ogawa et al. [37], respectively. The reaction mixtures for mammalian pols η, ι, and κ were the same as for calf pol α, and the reaction mixture for human pol λ was the same as for rat pol β. For pols, poly(dA)/oligo(dT)$_{18}$ (A/T = 2/1) and tritium-labeled 2'-deoxythymidine 5'-triphosphate ([^{3}H]-dTTP) were used as the DNA template-primer substrate and nucleotide [i.e., 2'-deoxynucleoside 5'-triphosphate (dNTP)] substrate, respectively.

The isolated compounds were dissolved in distilled dimethyl sulfoxide (DMSO) at various concentrations and sonicated for 30 sec. Aliquots of 4 μL of sonicated samples were mixed with 16 μL of each enzyme (final amount, 0.05 units) in 50 mM Tris-HCl (pH 7.5), containing 1 mM dithiothreitol, 50% glycerol, and 0.1 mM EDTA, and kept at 0°C for 10 min.

Table 1. IC$_{50}$ values of compounds 1–3 on the activities of various pols and other DNA metabolic enzymes

Enzyme	IC$_{50}$ values of compounds (µM)		
	1	2	3
--- Mammalian pols ---			
[Family A]			
Human pol γ	>100	>100	>100
[Family B]			
Calf pol α	>100	>100	>100
Human pol δ	>100	>100	>100
Human pol ε	>100	>100	>100
[Family X]			
Rat pol β	>100	>100	>100
Human pol λ	>100	9.1 ±0.6	12.2 ±0.8
[Family Y]			
Human pol η	16.1 ±0.8	>100	>100
Mouse pol ι	19.7 ±1.0	>100	>100
Human pol κ	10.2 ±0.5	>100	>100
--- Fish pols ---			
[Family B]			
Cherry salmon pol δ	>100	>100	>100
--- Insect pols ---			
[Family B]			
Fruit fly pol α	>100	>100	>100
Fruit fly pol δ	>100	>100	>100
Fruit fly pol ε	>100	>100	>100
--- Plant pols ---			
[Family B]			
Cauliflower pol α	>100	>100	>100
[Family X]			
Rice pol λ	>100	12.8 ±0.8	17.1 ±1.1
--- Prokaryotic pols ---			
E. coli pol I	>100	>100	>100
Taq pol	>100	>100	>100
T4 pol	>100	>100	>100
--- Other DNA metabolic enzymes ---			
T7 RNA polymerase	>100	>100	>100
Human DNA topoisomerase I	>100	>100	>100
Human DNA topoisomerase II	>100	>100	>100
T4 polynucleotide kinase	>100	>100	>100
Bovine deoxyribonuclease I	>100	>100	>100

Compounds incubated with each enzyme; enzyme activity in absence of compounds, 100%; and data, means ±SD (n=3).

These inhibitor-enzyme mixtures (8 μL) were added to 16 μL of each of the enzyme standard reaction mixtures and incubation carried out at 37°C for 60 min, except for *Taq* pol, which was incubated at 74°C for 60 min. Activity without the inhibitor was considered 100% and the remaining activity at each inhibitor concentration was determined relative to this value. One unit of pol activity was defined as the amount of enzyme that catalyzed incorporation of 1 nmol dNTP (i.e., dTTP) into synthetic DNA template-primers in 60 min at 37°C under the normal reaction conditions for each enzyme [35, 36].

4.3. Pol Inhibition by Compounds 1–3

Initially, the isolated compounds 1–3 from the waste gum fraction from crude soybean oil were investigated to determine whether they inhibited the activities of 9 mammalian pols. The purification grade of these compounds was more than 98%, determined by NMR analysis (data not shown).

Compound 1 was found to significantly inhibit the activities of human pol η, mouse pol ι, and human pol κ, all Y-family pols, which are DNA repair-related pols (Table 1). The 50% inhibitory concentration (IC_{50}) of compound 1 against these three pols was observed at concentrations of 16.1, 19.7, and 10.2 μM, respectively; therefore, the inhibitory effect on mammalian Y-family pols ranked as follows: pol κ > pol η > pol ι. Conversely, at high concentrations (i.e., 100 μM), this compound had no effect on the activities of other mammalian pol families, such as A-family of human pol γ, B-family of calf pol α, human pol δ and human pol ε, or X-family of rat pol β and human pol λ.

Compounds 2 and 3 inhibited the activity of human pol λ, with IC_{50} values of 9.1 and 12.2 μM, respectively; therefore, compound 2 was a stronger inhibitor than compound 3 by ~1.3-fold. These compounds had no influence at all on replicative pol activities, such as those from family B, human mitochondrial pol γ of family A, or repair-related pol activities, such as those from family X (pol β) and family Y. In particular, it was interesting that these compounds showed no effect even on pol β activity, which is thought to have a similar homology and three-dimensional structure to pol λ, although pols β and λ both belong to the X-family [4, 7]. Compounds 2 and 3 also inhibited plant (rice) pol λ activity to the same extent as they inhibited human pol λ.

In contrast, compounds 1–3 had no inhibitory effect on fish (i.e., cherry salmon) pol δ, insect (i.e., fruit fly) pols α, δ, and ε, plant (i.e., cauliflower) pol α, or prokaryotic pols, such as the Klenow fragment of *Escherichia coli* pol I, *Taq* pol, and T4 pol. When activated DNA (DNA with gaps digested by bovine deoxyribonuclease I) and dNTP were used as the DNA template-primer substrate and nucleotide substrate pair instead of synthesized DNA [poly(dA)/oligo(dT)$_{18}$ (A/T = 2/1)] and dTTP, respectively, the inhibitory effects of these compounds were unchanged (data not shown).

4.4. Mode of Inhibition of Eukaryotic Pol Species by Compounds 1–3

Next, to elucidate the mechanism of the selective inhibition of compounds 1–3 for eukaryotic pol species, we investigated the inhibitory mode of these compounds against human pols κ/λ. Poly(dA)/oligo(dT)$_{18}$ and dTTP were used as synthetic DNA template-primer substrate and nucleotide substrate, respectively, for kinetic analyses. The extent of inhibition as a function of the DNA template-primer substrate or nucleotide substrate concentration was measured (Table 2).

Double reciprocal plots (i.e., Lineweaver Burk plots) of the obtained data showed that the compound 1-induced inhibition of pol κ activity was noncompetitive with respect to both the DNA template-primer and nucleotide substrate. For the DNA template-primer, the apparent Michaelis constant (K_m) was unchanged at 1.54 μM, whereas decreases of 52.6, 29.3, 20.3, and 15.5 pmol/h in the maximum velocity (V_{max}) were observed in the presence of 0, 3, 6, and 9 μM compound 1, respectively. The K_m for the nucleotide substrate was unchanged at 2.00 μM and the V_{max} for the nucleotide substrate decreased from 41.7 to 15.7 pmol/h in the presence of 9 μM compound 1. Inhibition constant (K_i) values, obtained from Dixon plots, were found to be 3.0 and 4.2 μM for the DNA template-primer and nucleotide substrate, respectively. Because the K_i value for the DNA template-primer was ~1.4-fold smaller than that for the nucleotide substrate, the affinity of compound 1 was greater for the enzyme-DNA template-primer binary complex than for the enzyme-nucleotide substrate complex.

The collected data, expressed as double reciprocal plots, showed that compound 2 inhibited pol λ activity in a noncompetitive manner with respect to both the DNA template-primer substrate and nucleotide substrate. For the DNA template-primer substrate, the apparent K_m was unchanged at 2.38 μM compound 2, whereas 25.0, 40.2, and 49.9% decreases in V_{max} were observed in the presence of compound 2 at 2, 4, and 6 μM, respectively. The K_m for the nucleotide substrate was unchanged at 1.18 μM and the V_{max} for the nucleotide substrate decreased from 52.6 to 27.0 pmol/h in the presence of 0–6 μM compound 2. The K_i, obtained from Dixon plots, was found to be 5.3 μM for the DNA template-primer substrate and 4.5 μM for the nucleotide substrate.

Similarly, pol λ inhibition by compound 3 was noncompetitive with the DNA template-primer substrate as there was no change in the apparent K_m (2.38 μM), while the V_{max} decreased from 83.3 to 37.0 pmol/h for the DNA template-primer substrate in the presence of 0–9 μM compound 3. The induced inhibition of pol λ activity by compound 3 was noncompetitive with respect to the nucleotide substrate (K_m unchanged at 1.18 μM). The V_{max} for the nucleotide substrate was 2.10-fold less in the presence of 9 μM compound 3. From Dixon plots, the K_i value was 7.2 and 5.9 μM for the DNA template-primer and nucleotide substrates, respectively. As the K_i value for the nucleotide substrate was smaller than that for the DNA template-primer substrate, it was concluded that compounds 2 and 3 had greater affinity for the enzyme-nucleotide substrate binary complex than for the enzyme-DNA template-primer substrate complex.

When activated DNA and four dNTPs were used as the DNA template-primer substrate and nucleotide substrates, respectively, the mode of eukaryotic pol species inhibition by compounds 1–3 was the same as that with the above synthetic DNA template-primer (data not shown). These results suggested that these compounds might bind to or interact with a site distinct from both the DNA template-primer substrate binding site and the nucleotide substrate binding site of each pol.

**Table 2. Kinetic analysis of compounds 1–3 inhibitory effects
on the activity of human pol κ or human pol λ as a function of the DNA
template–primer substrate dose and the nucleotide substrate concentration**

Pol	Compound	Substrate	Compound conc. (μM)	K_m [a] (μM)	V_{max} [a] (pmol/h)	K_i [b] (μM)	Inhibitory mode
κ	1	DNA template-primer [c]	0	1.54	52.6	3.0	Noncompetitive
			3		29.3		
			6		20.3		
			9		15.5		
		dNTP [d]	0	2.00	41.7	4.2	Noncompetitive
			3		26.9		
			6		19.8		
			9		15.7		
λ	2	DNA template-primer [c]	0	2.38	83.3	5.3	Noncompetitive
			2		62.5		
			4		49.8		
			6		41.7		
		dNTP [d]	0	1.18	52.6	4.5	Noncompetitive
			2		40.0		
			4		32.3		
			6		27.0		
	3	DNA template-primer [c]	0	2.38	83.3	7.2	Noncompetitive
			3		58.8		
			6		45.4		
			9		37.0		
		dNTP [d]	0	1.18	52.6	5.9	Noncompetitive
			3		38.5		
			6		30.3		
			9		25.0		

[a] Data obtained from a Lineweaver Burk plot.
[b] Data obtained from a Dixon plot.
[c] Poly(dA)/oligo(dT)$_{18}$.
[d] dTTP.

5. EFFECTS OF COMPOUNDS 1–3 ON THE ACTIVITIES OF OTHER DNA METABOLIC ENZYMES

5.1. Other DNA Metabolic Enzymes Assays

Human DNA topoisomerases I and II, T7 RNA polymerase, T4 polynucleotide kinase, and bovine deoxyribonuclease I were prepared from commercial sources, and the activities of these enzymes measured in standard assays according to the manufacturer's specifications, as described by Mizushina et al. [38], Nakayama and Saneyoshi [39], Soltis and Uhlenbeck [40], and Lu and Sakaguchi [41], respectively.

5.2. Other DNA Metabolic Enzymes Inhibition by Compounds 1–3

Isolated compounds 1–3 barely influenced the activities of other DNA-metabolic enzymes, such as human DNA topoisomerases I and II, T7 RNA polymerase, T4 polynucleotide kinase, and bovine deoxyribonuclease I (Table 1). These results suggested that these compounds selectively inhibited the activities of mammalian Y-family of pols and eukaryotic pol λ. The fact that these glucosyl compounds, which are components of soybean, are inhibitors of a eukaryotic pol species is of great interest.

We next performed specific assays to determine whether compounds 1–3-induced inhibition resulted from the ability of these compounds to bind to DNA or to the enzyme. The interaction of compound 1, 2, or 3 with dsDNA was investigated in terms of the thermal transition of dsDNA. For this, the melting temperature (Tm) of dsDNA in the presence of an excess amount of each compound (100 µM) was observed using a spectrophotometer equipped with a thermoelectric cell holder. A thermal transition, indicated by a Tm, was not observed within the concentration range used in the assay, whereas a typical intercalating compound (ethidium bromide, 15 µM) as a positive control produced a clear thermal transition (data not shown).

The question of whether the inhibitory effect of these glucosyl compounds resulted from nonspecific adhesion to eukaryotic pol species or from its selective binding to specific sites was investigated by determining if an excessive amount of nucleic acid [poly(rC)] or protein (bovine serum albumin, BSA) prevented the inhibitory effect of compounds 1–3. Poly(rC) and BSA had little or no influence on pol inhibition by these compounds (data not shown), suggesting that compounds 1–3 selectively bound to the eukaryotic pol molecules. These observations indicated that these compounds did not act as DNA intercalating agents or as a template-primer substrate but, rather, that they bound directly to the enzyme and inhibited its activity.

Collectively, these results suggested that compounds 1–3 could be potent and selective inhibitors of eukaryotic pol species. We thus investigated whether pol inhibition by these compounds was effective for human cancer cell proliferation.

6. EFFECTS OF COMPOUNDS 1–3 ON CYTOTOXICITY

6.1. Cell Culture Method and Cell Viability Assay

The human cancer cell lines A549 (lung cancer), HCT116 (colon cancer), HeLa (cervical cancer), HepG2 (hepatocellular liver cancer), HL-60 (leukemia), MCF-7 (breast cancer), and NUGC-3 (stomach cancer) and normal cell lines HDF (human dermal fibroblasts) and HUVECs (human umbilical vein endothelial cells) were obtained from the American Type Culture Collection (Manassas, VA, USA). Human cancer cells were cultured in McCoy's 5A medium supplemented with 10% fetal bovine serum, 100 units/mL penicillin, and 100 µg/mL streptomycin. Normal human cells were cultured in Eagle's minimum essential medium supplemented with 4.5 g glucose/L plus 10% fetal calf serum, 5 mM L-glutamine, 50 units/mL penicillin, and 50 units/mL streptomycin. Cells were cultured at 37°C in a humidified incubator containing 5% CO_2/95% air. For assessment of cell growth, cells were plated at 1×10^4 cells per well in 96-well microplates, cultured for 12 h, and various concentrations of the isolated compounds added. The compounds were dissolved in DMSO to produce a 10-mM stock solution, which was diluted to the appropriate final concentrations with growth medium to 0.5% DMSO just before use. After cells were cultured with a compound, MTT [3-(4,5-dimethyl-2-thiazolyl)-2,5-diphenyl-2H tetrazolium bromide] solution was added, and the mixture incubated for 3 h. The number of viable cells in each well was then counted using a microplate reader at 570 nm (MTT assay) [42].

6.2. Cytotoxicity by Compounds 1–3

Pols have recently emerged as important cellular targets for chemical intervention in the development of anticancer agents. As compounds 1–3 could thus be useful in chemotherapy, we investigated the cytotoxic effect of these compounds against cultured human cancer and normal cell lines. As shown in Table 2, compounds 1–3 treatment for 24 h did not suppress the growth of seven human cancer cell lines, and treatment for 72 h also had no detectable cytotoxicity in human normal cells, such as HDF and HUVECs. Aphidicolin, which is a DNA replicative pols α, δ, and ε inhibitor, suppressed human cancer and normal cell growth with 50% lethal dose (LD_{50}) of approximately 20 μM, halted cell cycle in S phase, and induced apoptosis [43]. These results suggested that aphidicolin might be able to penetrate cancer cells and reach the nucleus, inhibiting DNA replicative pol activities, and such inhibition might lead to cell death. In contrast, compounds 1–3 did not inhibit the activities of pols α, δ, and ε; therefore, these compounds might not influence cell proliferation.

7. EFFECTS OF COMPOUNDS 1–3
ON ANTI-INFLAMMATORY ACTIVITY

7.1. Anti-inflammatory Assay by TPA-induced Inflammation in the Mouse

A mouse inflammatory test was performed according to Gschwendt's method [44]. In brief, an acetone solution of one of the three compounds (125, 250, or 500 μg in 20 μL) or 20 μL of acetone as a vehicle control was applied to the inner part of a mouse ear. Thirty min after test compound application, a TPA solution (0.5 μg/20 μL of acetone) was applied to the same site on the ear. TPA solution was applied as a control to the other ear of the same mouse. After 7 h, a punch biopsy disk (6 mm diameter) was obtained from each ear and weighed. The inhibitory effect (IE) of the test solution was expressed as a ratio of the test solution-prevented increase in weight to that of the TPA only response of the ear disks:

IE = [(TPA only) − (tested compound plus TPA)] / [(TPA only) − (vehicle)] × 100.

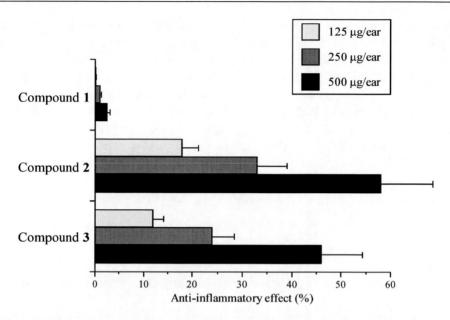

Figure 3. Anti-inflammatory activity of compounds 1–3 toward TPA-induced edema on mouse ear. Each compound (125, 250, and 500 μg) applied individually to one ear of a mouse, and after 30 min TPA (0.5 μg) applied to both ears; edema evaluated after 7 h; inhibitory effect expressed as percentage of edema; and data, means ±SD (n=6).

This animal study was performed according to the guidelines outlined in the "Care and Use of Laboratory Animals" of Kobe Gakuin University. The animals were anesthetized with pentobarbital before undergoing cervical dislocation. Female 8-week-old ICR mice bred in-house were used for all experiments. All mice were maintained under a 12-h light/dark cycle and housed with food and water *ad libitum* at a room temperature of 25°C.

7.2. Anti-inflammation by Compounds 1–3

In a previous pol inhibitor study, we found that there was a relationship between pol λ inhibitors and TPA-induced anti-inflammatory activity [8, 9, 13, 14]. Thus, using the mouse ear inflammatory test *in vivo*, we examined the anti-inflammatory activity of compounds 1–3. Application of TPA (0.5 μg) to the mouse ear induced edema, resulting in a 241% increase in the weight of the ear disk by 7 h after application. As shown in Figure 3, compound 1 exhibited no anti-inflammatory activity, but pretreatment with compounds 2 and 3 dose-dependently suppressed inflammation. The effect of compound 2 was stronger than that of compound 3 by ~1.3-fold. Therefore, the anti-inflammatory effect *in vivo* of these three compounds displayed the same order as their inhibitory effect *in vitro* on pol λ (Table 1). These results suggested that inhibition of pol λ activity correlated positively with the anti-inflammatory activity observed.

Human cancer cell lines A549, HCT116, HeLa, HepG2, HL-60, MCF-7, and NUGC-3, and normal human cell lines HDF and HUVECs incubated with each compound for 24 and 72 h, respectively; cell viability determined using MTT assay [43]; data, means ±SD (n=5).

**Table 3. Inhibitory effect of compounds 1–3 on human cancer
and normal cell proliferation**

	Cell line	Cell type	LD_{50} values of compounds (µM)		
			1	2	3
Cancer cells	A549	Lung cancer	>200	>200	>200
	HCT116	Colon cancer	>200	>200	>200
	HeLa	Cervix cancer	>200	>200	>200
	HepG2	Liver cancer	>200	>200	>200
	HL-60	Leukemia	>200	>200	>200
	MCF-7	Breast cancer	>200	>200	>200
	NUGC-3	Stomach cancer	>200	>200	>200
Normal cells	HDF	Dermal fibroblast from skim	>200	>200	>200
	HUVEC	Umbilical vein endothelial cell	>200	>200	>200

8. DISCUSSION

8.1. Compound 1 as a Selective Inhibitor of Mammalian Y-family Pols

As described in this review, we isolated compound 1 as a potent inhibitor specific to mammalian Y-family pols from the waste gum fraction of crude soybean oil. This compound was an acylated steryl glycoside, β-sitosteryl (6'-O-linoleoyl)-glucoside (Figure 1A). The structure of acylated steryl glycoside comprised the 3β-hydroxyl of the sterol moiety linked to the hydroxyl C_1 position of the sugar, and the fatty acid esterified at position C_6 of the sugar. Compound 1 consisted of a conjugated structure of β-sitosterol, linoleic acid, and D-glucose as a sterol, a fatty acid, and a sugar, respectively (Figure 1A). We previously reported that some fatty acids inhibit eukaryotic pol activity [3, 35, 36] and, here, compound 1 showed stronger effects in inhibiting Y-family pols than linoleic acid. Thus, the molecular structure of compound 1 may be important for the inhibition of Y-family pol activities. More detailed examination of the mechanism of selective inhibition for eukaryotic pol species by compound 1 will be addressed in further studies.

Eukaryotic cells reportedly contain 14 pol species belonging to four families: namely, family A (pols γ, θ, and ν), family B (pols α, δ, ε, and ζ), family X (pols β, λ, and µ), and family Y (pols η, ι, and κ, and REV1) [6, 45]. Y-family pols differ from pols belonging to other families in their ability to replicate through damaged DNA. Members of this family are hence called translesion synthesis (TLS) pols [46]. Depending on the lesion, TLS pols can bypass the damage in an error-free or error-prone fashion, the latter resulting in elevated mutagenesis. Xeroderma pigmentosum variant (XPV) patients, for instance, have mutations in the gene encoding pol η, which is able to bypass cyclobutane pyrimidine dimers, the most frequent UV-induced lesions, in an error-free fashion. In XPV patients, alternative error-prone pols, e.g., pol ζ (a B-family pol), are thought to be involved in errors that result in the cancer predisposition of these patients. Other members of the Y-family in humans (pols ι, κ, and REV1) are thought to be involved in bypassing different lesions [45, 46]. Inhibitors of Y-family pols may be useful as anticancer drugs for clinical radiation therapy. Acylated steryl glycosides, such as compound 1, and their constituent parts, such as β-sitosterol, linoleic acid and D-glucose, are common components of plant membranes [47]. Whether acylated steryl glycosides are active metabolic compounds is unknown. Some hypotheses consider these conjugates to be final products of sterol metabolism, whereas others suggest that the processes of deacylation of acylated steryl glycosides incorporated into membranes can play significant roles in membrane properties or in the modulation of membranous enzyme activity. Some acylated steryl glycosides are found in sprout-derived sterols of plants [48, 49] and play important roles as sprout-growing biochemical factors that are newly generated during the process of germination. Several biological effects of acylated steryl glycoside are known, including not only anti-inflammatory [50] but also antiulcerogenic activity [51, 52], and antidiabetic [53] effects. The possible bioactivities by compound 1 detected here might be related to the selective inhibition of Y-family pol activities.

8.2. Compounds 2 and 3 as Eukaryotic Pol λ Specific Inhibitors and Anti-Inflammation

Compounds 2 and 3 were isolated as a steroidal glycoside (eleutheroside A, Figure 2B) and a cerebroside (glucosyl ceramide, AS-1-4, Figure 2C), respectively, and specifically inhibited eukaryotic pol λ activity. Among the X family of pols, pol λ appears to work in a similar manner to pol β [54]. Pol β is involved in the short-patch base excision repair (BER) pathway [55–58] as well as playing an essential role in neural development [59]. Recently, pol λ has been found to possess 5'-deoxyribose-5-phosphate lyase activity but not apurinic/apyrimidinic lyase activity [60]. Pol λ is able to substitute for pol β during BER *in vitro*, suggesting that pol λ also participates in BER. Northern blot analysis has indicated that transcripts of pol β are abundantly expressed in the testis, thymus, and brain in rats [61], whereas pol λ is efficiently transcribed mostly in the testis [62]. Bertocci et al. reported that mice in which pol λ expression is knocked out are not only viable and fertile but also display a normal hypermutation pattern [63]. As well as causing inflammation, TPA influences cell proliferation and has physiological effects on cells because of its tumor promoter activity [64]. Therefore, anti-inflammatory agents are expected to suppress DNA replication/ repair/recombination in nuclei in relation to the action of TPA. As pol λ is a repair/recombination-related pol [54], our finding—that the molecular target of compounds 2 and 3 was pol λ—is in good agreement with an expected mechanism where anti-inflammatory agents suppress DNA repair/recombination. The detailed mechanism by which these compounds prevent mammalian pol λ inhibition and, hence, could inhibit inflammation remains unclear. Thus, further studies will be conducted to clarify the exact mechanism of the anti-inflammatory effect of these glucosyl compounds.

CONCLUSION

In this review, we screened for pol inhibitors isolated from the waste gum fraction produced in the degumming process during soybean oil refining, resulting in the purification and identification of compounds 1–3, which potently and selectively inhibited the activities of eukaryotic pol species. These three compounds are all conjugated with one glucose molecule (Figure 2) and, therefore, this glucosyl moiety might be important for their bioactivities, such as their inhibitory effects on the activity of eukaryotic pol species and inflammation. These glucosyl compounds could be useful molecular tools as pol-specific inhibitors in studies to determine the precise roles of each pol and/or family of pols *in vitro*.

In soybean, protein and oil account for ~60% of dry soybeans by weight, at 40 and 20%, respectively [65]. The remainder consists of 35% carbohydrate and ~5% ash. Soybean cultivars comprise ~8% seed coat or hull, 90% cotyledons, and 2% hypocotyl axis or germ. Thus, purified glucosyl compounds, especially compounds 2 and 3, and/or the purified fraction, containing large amounts of these compounds, could be used as an anti-inflammatory functional food and/or cosmetic based on the ability to inhibit eukaryotic pol λ activity. These compounds may also be effective nutrients for human anti-inflammatory health promotion.

CONFLICT OF INTEREST

None of the authors have any financial interest in any of the compounds reviewed in this article.

ACKNOWLEDGMENTS

We are grateful for the donations of calf pol α from Dr. M. Takemura of Tokyo University of Science (Tokyo, Japan), rat pol β, human pols δ and ε from Dr. K. Sakaguchi of Tokyo University of Science (Chiba, Japan), human pol γ from Dr. M. Suzuki of Nagoya University School of Medicine (Nagoya, Japan), human pols η and ι from Dr. F. Hanaoka and Dr. C. Masutani of Osaka University (Osaka, Japan), human pol κ from Dr. H. Ohmori of Kyoto University (Kyoto, Japan), and human pol λ from Dr. O. Koiwai of Tokyo University of Science (Chiba, Japan).

This study was supported in part by MEXT (Ministry of Education, Culture, Sports, Science and Technology, Japan)-Supported Program for the Strategic Research Foundation at Private Universities, 2012–2016. Y.M. acknowledges Grants-in-Aid for Scientific Research (C) (No. 24580205) from MEXT, the Takeda Science Foundation (Japan) and the Nakashima Foundation (Japan). I. K. acknowledges a Grant-in-Aid for Young Scientists (B) (No. 23710262) from MEXT.

REFERENCES

[1]   Kornberg, A. and Baker, T.A. (1992). Eukaryotic DNA polymerase, "DNA replication" 2nd edition. Freeman, W.D., and Co., New York, Chapter 6, pp. 197-225.

[2]   DePamphilis, M.L. (1996). DNA replication in eukaryotic cells. Cold Spring Harbor Laboratory Press, in U.S.A.

[3]   Mizushina, Y., Yagi, H., Tanaka, N., Kurosawa, T., Seto, H., Katsumi, K., Onoue, M., Ishida, H., Iseki, A., Nara, T., Morohashi, K., Horie, T., Onomura, Y., Narusawa, M., Aoyagi, N., Takami, K., Yamaoka, M., Inoue, Y., Matsukage, A., Yoshida, S. and Sakaguchi, K. (1996). Screening of inhibitor of eukaryotic DNA polymerases produced by microorganisms. *J. Antibiot. (Tokyo)*, *49*, 491-492.

[4]   Hubscher, U., Maga, G. and Spadari, S. (2002). Eukaryotic DNA polymerases. *Annu. Rev. Biochem.*, *71*, 133-163.

[5]   Bebenek, K. and Kunkel, T.A. (2004). DNA Repair and Replication, "Advances in Protein Chemistry", Elsevier, San Diego, Yang, W. (Ed.), vol. 69, pp. 137-165.

[6]   Takata, K., Shimizu, T., Iwai, S. and Wood R.D. (2006). Human DNA polymerase N (POLN) is a low fidelity enzyme capable of error-free bypass of 5S-thymine glycol. *J. Biol. Chem.*, *281*, 23445-23455.

[7]   Loeb, L.A. and Monnat, Jr. R.J. (2008). DNA polymerases and human disease. *Nat. Rev. Genet.*, *9*, 594-604.

[8]   Mizushina, Y. (2009). Specific inhibitors of mammalian DNA polymerase species. *Biosci. Biotechnol. Biochem.*, *73*, 1239-1251.

[9]   Mizushina, Y. (2011). Screening of novel bioactive compounds from food components and nutrients. *J. Jpn. Soc. Nutr. Food Sci.*, *64*, 377-384.

[10] Mizushina, Y., Kamisuki, S., Kasai, N., Ishidoh, T., Shimazaki, N., Takemura, M., Asahara, H., Linn, S., Yoshida, S., Koiwai, O., Sugawara, F., Yoshida, H. and Sakaguchi, K. (2002). Petasiphenol: a DNA polymerase λ inhibitor. *Biochemistry*, *41*, 14463-14471.

[11] Mizushina, Y., Ishidoh, T., Takeuchi, T., Shimazaki, N., Koiwai, O., Kuramochi, K., Kobayashi, S., Sugawara, F., Sakaguchi, K. and Yoshida, H. (2005). Monoacetylcurcumin: a new inhibitor of eukaryotic DNA polymerase λ and a new ligand for inhibitor-affinity chromatography. *Biochem. Biophys. Res. Commun.*, *337*, 1288-1295.

[12] Takeuchi, T., Ishidoh, T., Iijima, H., Kuriyama, I., Shimazaki, N., Koiwai, O., Kuramochi, K., Kobayashi, S., Sugawara, F., Sakaguchi, K., Yoshida, H. and Mizushina, Y. (2006). Structural relationship of curcumin derivatives binding to the BRCT domain of human DNA polymerase λ. *Genes Cells*, *11*, 223-235.

[13] Mizushina, Y., Hirota, M., Murakami, C., Ishidoh, T., Kamisuki, S., Shimazaki, N., Takemura, M., Perpelescu, M., Suzuki, M., Yoshida, H., Sugawara, F., Koiwai, O. and Sakaguchi, K. (2003). Some anti-chronic inflammatory compounds are DNA polymerase λ-specific inhibitors. *Biochem. Pharmacol.*, *66*, 1935-1944.

[14] Mizushina, Y., Takeuchi, T., Kuramochi, K., Kobayashi, S., Sugawara, F., Sakaguchi, K. and Yoshida, H. (2007). Study on the molecular structure and bio-activity (DNA polymerase inhibitory activity, anti-inflammatory activity and anti-oxidant activity) relationship of curcumin derivatives. *Curr. Bioactive Compounds*, *3*, 171-177.

[15] Nishida, M., Nishiumi, S., Mizushina, Y., Fujishima, Y., Yamamoto, K., Masuda, A., Mizuno, S., Fujita, T., Morita, Y., Kutsumi, H., Yoshida, H., Azuma, T. and Yoshida, M. (2010). Monoacetylcurcumin strongly regulates inflammatory responses through inhibition of NF-κB activation. *Int. J. Mol. Med.*, *25*, 761-767.

[16] Karasulu, H.Y., Karasulu, E., Büyükhelvacıgil, M., Yıldız, M., Ertugrul, A., Büyükhelvacıgil, K., Ustün, Z. and Gazel, N. (2011). Soybean oil: Production process,

benefits and uses in pharmaceutical dosage form. "Soybean and Health", In Tech, El-Shemy, H. (Ed.), Chapter 13, pp. 283-310.

[17] Mahadevappa, V.G. and Raina, P.L. (1981). Sterols, esterified sterols, and glycosylated sterols of cow pea lipids (*Vigna uguiculata*). *J. Agric. Food Chem.*, 29, 1225-1227.

[18] Yamauchi, R., Aizawa, K., Inakuma, T. and Kato, K. (2001). Analysis of molecular species of glycolipids in fruit pastes of red bell pepper (*Capsicum annuum* L.) by high-performance liquid chromatography-mass spectrometry. *J. Agric. Food Chem.*, 49, 622-627.

[19] Chung, I.M., Ali, M., Ahmad, A., Chun, S.C., Kim, J.T., Sultana, S., Kim, J.S., Min, S.K and Seo, B.R. (2007). Steroidal constituents of rice (*Rryza sativa*) hulls with algicidal and herbicidal activity against blue-green algae and duckweed. *Phytochem. Anal.*, 18, 133-145.

[20] Inagaki, M., Harada, Y., Yamada, K., Isobe, R., Higuchi, R., Matsuura, H. and Itakura, Y. (1998). Isolation and structure determination of cerebrosides from garlic, the bulbs of *Allium sativum* L. *Chem. Pharm. Bull.*, 46, 1153-1156.

[21] Faizi, S., Ali, M., Saleem, R., Irfanullah and Bibi, S. (2001). Complete ^{1}H and ^{13}C NMR assignments of stigma-5-en-3-*O*-β-glucoside and its acetylderivative. *Magn. Reson. Chem.*, 39, 399-405.

[22] Cateni, F., Zilic, J. and Zacchigna, M. (2008). Isolation and Structure elucidation of cerebrosides from *Euphorbia platyphyllos* L.. *Sci. Pharm.*, 76, 451-469.

[23] Tamai, K., Kojima, K., Hanaichi, T., Masaki, S., Suzuki, M., Umekawa, H. and Yoshida, S. (1998). Structural study of immunoaffinity-purified DNA polymerase α-DNA primase complex from calf thymus. *Biochim. Biophys. Acta*, 950, 263-273.

[24] Date, T., Yamaguchi, M., Hirose, F., Nishimoto, Y., Tanihara, K. and Matsukage, A. (1998). Expression of active rat DNA polymerase β in *Escherichia coli*. *Biochemistry*, 27, 2983-2990.

[25] Umeda, S., Muta, T., Ohsato, T., Takamatsu, C., Hamasaki, N. and Kang, D. (2000). The D-loop structure of human mtDNA is destabilized directly by 1-methyl-4-phenylpyridinium ion (MPP+), a parkinsonism-causing toxin. *Eur. J. Biochem.*, 267, 200-206.

[26] Oshige, M., Takeuchi, R., Ruike, R., Kuroda, K. and Sakaguchi, K. (2004). Subunit protein-affinity isolation of *Drosophila* DNA polymerase catalytic subunit. *Protein Expr. Purif.*, 35, 248-256.

[27] Kusumoto, R., Masutani, C., Shimmyo, S., Iwai, S. and Hanaoka, F. (2004). DNA binding properties of human DNA polymerase η: implications for fidelity and polymerase switching of translesion synthesis. *Genes Cells*, 9, 1139-1150.

[28] Ohashi, E., Murakumo, Y., Kanjo, N., Akagi, J., Masutani, C., Hanaoka, F. and Ohmori, H. (2004). Interaction of hREV1 with three human Y-family DNA polymerases. *Genes Cells*, 9, 523-531.

[29] Shimazaki, N., Yoshida, K., Kobayashi, T., Toji, S., Tamai, T. and Koiwai, O. (2000). Over-expression of human DNA polymerase λ in *E. coli* and characterization of the recombinant enzyme. *Genes Cells*, 7, 639–651.

[30] Yamaguchi, T., Saneyoshi, M., Takahashi, H., Hirokawa, S., Amano, R., Liu, X., Inomata, M. and Maruyama, T. (2006). Synthetic Nucleoside and Nucleotides. 43. Inhibition of vertebrate telomerases by carbocyclic oxetanocin G (C.OXT-G)

triphosphate analogues and influence of C.OXT-G treatment on telomere length in human HL60 cells. *Nucleos. Nucleot. Nucl.*, *25*, 539-551.

[31]   Aoyagi, N., Matsuoka, S., Furunobu, A., Matsukage, A. and Sakaguchi, K. (1994). *Drosophila* DNA polymerase δ: Purification and characterization. *J. Biol. Chem.*, *269*, 6045-6050.

[32]   Aoyagi, N., Oshige, M., Hirose, F., Kuroda, K., Matsukage, A. and Sakaguchi, K. (1997). DNA polymerase ε from *Drosophila melanogaster*. *Biochem. Biophys. Res. Commun.*, *230*, 297-301.

[33]   Sakaguchi, K., Hotta, Y. and Stern, H. (1980). Chromatin-associated DNA polymerase activity in meiotic cells of lily and mouse. *Cell Struct. Funct.*, *5*, 323-334.

[34]   Uchiyama, Y., Kimura, S., Yamamoto, T., Ishibashi, T. and Sakaguchi, K. (2004). Plant DNA polymerase λ, a DNA repair enzyme that functions in plant meristematic and meiotic tissues. *Eur. J. Biochem.*, *271*, 2799-2807.

[35]   Mizushina, Y., Tanaka, N., Yagi, H., Kurosawa, T., Onoue, M., Seto, H., Horie, T., Aoyagi, N., Yamaoka, M., Matsukage, A., Yoshida, S. and Sakaguchi, K. (1996). Fatty acids selectively inhibit eukaryotic DNA polymerase activities in vitro. *Biochim. Biophys. Acta*, *1308*, 256-262.

[36]   Mizushina, Y., Yoshida, S., Matsukage, A. and Sakaguchi, K. (1997). The inhibitory action of fatty acids on DNA polymerase β. *Biochim. Biophys. Acta*, *1336*, 509-521.

[37]   Ogawa, A., Murate, T., Suzuki, M., Nimura, Y. and Yoshida, S. (1998). Lithocholic acid, a putative tumor promoter, inhibits mammalian DNA polymerase β. *Jpn. J. Cancer Res.*, *89*, 1154-1159.

[38]   Mizushina, Y., Nishimura, K., Takenaka, Y., Takeuchi, T., Sugawara, F., Yoshida, H. and Tanahashi, T. (2010). Inhibitory effects of docosyl p-coumarate on DNA topoisomerase activity and human cancer cell growth. *Int. J. Oncol.*, *37*, 993-1000.

[39]   Nakayama, C. and Saneyoshi, M. (1985). Inhibitory effects of 9-β-D-xylofuranosyladenine 5'-triphosphate on DNA-dependent RNA polymerase I and II from cherry salmon (*Oncorhynchus masou*). *J. Biochem. (Tokyo)*, *97*, 1385-1389.

[40]   Soltis, D.A. and Uhlenbeck, O.C. (1982). Isolation and characterization of two mutant forms of T4 polynucleotide kinase. *J. Biol. Chem.*, *257*, 11332-11339.

[41]   Lu, B.C. and Sakaguchi, K. (1991). An endo-exonuclease from meiotic tissues of the basidiomycete *Coprinus cinereus*: Its purification and characterization. *J. Biol. Chem.*, *266*, 21060-21066.

[42]   Mosmann, T. (1983). Rapid colorimetric assay for cellular growth and survival: application to proliferation and cytotoxicity assays. *J. Immunol. Methods*, *65*, 55-63.

[43]   Kuriyama, I., Mizuno, T., Fukudome, K., Kuramochi, K., Tsubaki, K., Usui, T., Imamoto, N., Sakaguchi, K., Sugawara, F., Yoshida, H. and Mizushina, Y. (2008). Effect of dehydroaltenusin-C12 derivative, a selective DNA polymerase alpha inhibitor, on DNA replication in cultured cells. *Molecules*, *13*, 2948-2961.

[44]   Gschwendt, M., Kittstein, W., Furstenberger, G. and Marks, F. (1984). The mouse ear edema: a quantitatively evaluable assay for tumor promoting compounds and for inhibitors of tumor promotion. *Cancer Lett.*, *25*, 177-185.

[45]   Friedberg, E.C., Feaver, W.J. and Gerlach, V.L. (2000). The many faces of DNA polymerases: strategies for mutagenesis and for mutational avoidance. *Proc. Natl. Acad. Sci. USA*, *97*, 5681-5683.

[46] Prakash, S., Johnson, R.E. and Prakash, L. (2005). Eukaryotic translesion synthesis DNA polymerases: specificity of structure and function. *Annu. Rev. Biochem.*, *74*, 317-353.

[47] Wojciechowski, Z.A. (1991). Biochemistry of phytosterols conjugates. "Physiology and Biochemistry of Sterols", AOCS press, IL, USA, Patterson, G.W. and Nes, W.D. (Ed.), pp. 361-393.

[48] Pegel, K.H. (1997). The importance of sitosterol and sitosterolin in human and animal nutrition. *S. Afr. J. Sci.*, *93*, 263-268.

[49] Zhang, H., Vasanthan, T. and Wettasinghe, M. (2007). Enrichment of tocopherols and phytosterols in canola oil during seed germination. *J. Agric. Food Chem.*, *55*, 355-359.

[50] Gupta, M.B., Nath, R., Srivastava, N., Shankar, K., Kishor, K. and Bhargava, K.P. (1980). Anti-inflammatory and antipyretic activities of β-sitosterol. *Planta Medica, 39*, 157-163.

[51] Okuyama, E. and Yamazaki, M. (1983). The principles of Tetragonia tetragonoides having an antiulcerogenic activity. I. Isolation and identification of sterylglucoside mixture (compound A). *Yakugaku Zasshi, 103*, 43-48.

[52] Xiao, M., Yang, Z., Jiu, M., You, J. and Xiao, R. (1992). The antigastroulcerative activity of β-sitosterol- β-D-glucoside and its aglycone in rats. *Hua Xi Yi Ke Da Xue Xue Bao, 23*, 98-101.

[53] Ivorra, M.D., D'Ocon, M.P., Paya, M. and Villar, A. (1988). Antihyperglycemic and insulin-releasing effects of β-sitosterol 3-β-Dglucoside and its aglycone, β-sitosterol. *Arch. Int. Pharmacodyn. Ther.*, *296*, 224-231.

[54] Garcia-Diaz, M., Bebenek, K., Sabariegos, R., Dominguez, O., Rodriguez, J., Kirchhoff, T., Garcia-Palomero, E., Picher, A.J., Juarez, R., Ruiz, J.F., Kunkel, T.A. and Blanco, L. (2002). DNA polymerase λ, a novel DNA repair enzyme in human cells. *J. Biol. Chem.*, *277*, 13184-13191.

[55] Singhal, R.K. and Wilson, S.H. (1993). Short gap-filling synthesis by DNA polymerase β is processive. *J. Biol. Chem.*, *268*, 15906-15911.

[56] Matsumoto, Y. and Kim, K. (1995). Excision of deoxyribose phosphate residues by DNA polymerase β during DNA repair. *Science, 269*, 699-702.

[57] Sobol, R.W., Horton, J.K., Kuhn, R., Gu, H., Singhal, R.K., Prasad, R., Rajewsky, K. and Wilson, S.H. (1996). Requirement of mammalian DNA polymerase-β in base-excision repair. *Nature, 79*, 183-186.

[58] Ramadan, K., Shevelev, I.V., Maga, G. and Hubscher, U. (2002). DNA polymerase λ from calf thymus preferentially replicates damaged DNA. *J. Biol. Chem.*, *277*, 18454-18458.

[59] Sugo, N., Aratani, Y., Nagashima, Y., Kubota, Y. and Koyama, H. (2000). Neonatal lethality with abnormal neurogenesis in mice deficient in DNA polymerase β. *EMBO J.*, *19*, 1397-1404.

[60] Garcia-Diaz, M., Bebenek, K., Kunkel, T.A. and Blanco, L. (2001). Identification of an intrinsic 5'-deoxyribose-5-phosphate lyase activity in human DNA polymerase λ: a possible role in base excision repair. *J. Biol. Chem.*, *276*, 34659-34663.

[61] Hirose, F., Hotta, Y., Yamaguchi, M. and Matsukage, A. (1989). Difference in the expression level of DNA polymerase β among mouse tissues: high expression in the pachytene spermatocyte. *Exp. Cell Res.*, *181*, 169-180.

[62] Garcia-Diaz, M., Dominguez, O., Lopez-Fernandez, L.A., De Lera, L.T., Saniger, M.L., Ruiz, J.F., Parraga, M., Garcia-Ortiz, M.J., Kirchhoff, T., Del Mazo, J., Bernad, A. and Blanco, L. (2000). DNA polymerase λ, a novel DNA repair enzyme in human cells. *J. Mol. Biol.*, *301*, 851-867.

[63] Bertocci, B., De Smet, A., Flatter, E., Dahan, A., Bories, J.C., Landreau, C., Weill, J.C. and Reynaud, C.A. (2002). Cutting edge: DNA polymerases μ and λ are dispensable for Ig gene hypermutation. *J. Immunol.*, *168*, 3702-3706.

[64] Nakamura, Y., Murakami, A., Ohto, Y., Torikai, K., Tanaka, T. and Ohigashi, H. (1995). Suppression of tumor promoter-induced oxidative stress and inflammatory responses in mouse skin by a superoxide generation inhibitor 1'-acetoxychavicol acetate. *Cancer Res.*, *58*, 4832-4839.

[65] Singh, R.J., Nelson, R.L. and Chung, G. (2007). Soybean (*Glycine max* (L.) Merr.). "Genetic resources, chromosome engineering, and crop improvement", CRC Press, NY, USA, Singh, R.J. (Ed.), pp. 13-50.

In: Agricultural Research Updates. Volume 5
Editors: P. Gorawala and S. Mandhatri

ISBN: 978-1-62618-723-8
© 2013 Nova Science Publishers, Inc.

Chapter 4

DRY MATTER PRODUCTION, YIELD DYNAMICS AND CHEMICAL COMPOSITION OF PERENNIAL GRASS AND FORAGE LEGUME MIXTURES AT VARIOUS SEEDING RATE PROPORTIONS

Tessema Zewdu Kelkay[*]

School of Animal and Sciences, College of Agriculture and Environmental Sciences,
Haramaya University, Dire Dawa, Ethiopia

ABSTRACT

Dry matter (DM) yield, relative yield, relative total yield, aggresivity index, relative crowding coefficient and chemical compositions of grass-legume mixtures were studied in a field experiment at Haramaya University field station, Ethiopia in 2005 and 2006. *Chloris gayana, Panicum coloratum, Melilotus alba* and *Medicago sativa* were planted as pure stands and in mixtures using 50:50, 67:33, 33:67, 75:25 and 25:75 seed rate proportions of grasses and legumes, respectively. *C. gayana* mixed with *M. alba* at seed rates of 50:50 and 33:67 produced DM yields of 26.7 and 26.1 t ha^{-1}, respectively, which were higher than other mixtures and pure stands. The total yields of all grass-legume mixtures were greater than monocultures. The mean relative crowding coefficient values of the grasses in the mixtures with legume species were high, indicating that they contributes to higher DM yields of the grass-legume combinations. Forage legume monoculture and legume-grass mixtures had higher crude protein contents than pure grass stand, whereas grass monocultures had higher fibre fraction contents compared to legume monocultures and legume-grass mixtures. *C. gayana* mixed with *M. alba* at seeding rate of 50:50 and 33:67 can be introduced to alleviate the feed shortages into smallholder farms in the eastern parts of Ethiopia. Further studies on animal performance should be conducted using feeding experiments and under grass-legume mixed pasture grazing to confirm the performance of these mixtures under farmers conditions.

---

[*] Email: tessemaz@yahoo.com.

Keywords: Crop-livestock mixed farming; Crude protein; Ethiopia; Grass-legume mixture; Feed shortage; Fibre fraction; Nutritive value

INTRODUCTION

Both under nutrition and malnutrition are the major constraints for livestock production in all parts of Ethiopia throughout the year (Mekoya, 2008; Yeshitila et al., 2008a, b). The major feed resources of the country are natural pasture and crop residues, and these feedstuffs are inadequate to satisfy the nutritional requirements of ruminants (Mekoya, 2008). However, the development of grass-legume mixed pasture is one of the alternative strategies to enhance the availability of feed resources in terms of quantity and quality. Forage quality and seasonal distribution of dry matter (DM) yield of grass-legume mixed pasture has been proved to be superior to those of grasses or legumes grown alone (Minson, 1990; Daniel, 1996; Diriba et al., 2002; Tessema and Baars, 2006). Moreover, the emphasis under smallholder crop-livestock production systems should focus on low-input and sustainable strategies of feed development. Accordingly, grass-legume mixture based forage development provides better land use efficiency than the respective monoculture without an additional investment (Prasad, 1981; Tessema, 1996; Berhan, 2006; Tessema and Baars, 2006). Therefore a grass-legume mixed pasture could play an important role in improving both the quantity and quality of forage without an additional organic and/or inorganic fertilizer applications compared to the monocultures. Moreover, a grass-legume mixed pasture provides advantages over monocultures; it reduces bloat from forage legumes and the incidence of crop diseases and insect pests as well as improves soil erosion (Mannteje, 1984; Minson, 1990).

The adaptability and DM production of promising perennial grasses and herbaceous forage legumes were evaluated in pure stands in the eastern parts of Ethiopia in the past years (Berhan, 2005, 2006). However, the DM yield performance and nutritional quality of grass-legume mixed pasture largely depends on the compatibility of grass and legume species and the relative seed proportion of the grass and legume components in the mixture. This is because of the fact that a low seed rate in the mixture may result in poor stand and low DM yield, whereas a high seed rates in the grass-legume mixed pasture may not be economical for smallholder livestock producers due to the high cost of forage seeds in developing countries (Prasad, 1981; Larbi et al., 1995; Tessema, 1996; Berhan, 2006; Tessema and Baars, 2006). According to Lema et al. (1991) and Diriba (2002), a progressive increase in the contribution of the legume component to the total DM yield compared to the grass component as the seed rate of the legume increases in the grass-legume mixed pasture and vice-versa has been reported in sub-humid western parts of Ethiopia. Combining both the DM yield and nutritional quality of feeds under smallholder livestock production systems in Ethiopia could be possible by growing grass-legume mixed pastures with combatable grass and legume species and appropriate seed rate proportion of each component in the mixture (Larbi et al., 1995; Tessema, 1996; Tessema and Baars, 2004, 2006). The objective of this study, therefore, was to evaluate the DM yield production, biological potential and chemical compositions (nutritional quality) of different pure stand grass and herbaceous forage legume species and their mixed pastures at various seed rate proportions at Haramaya University in the eastern part of Ethiopia.

MATERIALS AND METHODS

Description of Study Area

A grass-legume experiment was conducted during 2004 and 2005 at Haramaya University Research Centre (9° 26' N, 42° 03' E; 1980 masl), 511 km from Addis Ababa on alluvial-vertisols (Tamire, 1982, Berhan, 2005, 2006). The 0-40cm layer of the soil before sowing and fertiliser application had a pH of 6.34, total N of 0.16, available phosphorus of 0.66 ppm, organic matter of 2.28% and organic carbon of 1.33% (Tessema 2008). The twenty years mean annual rainfall of the area is 625 mm and the average annual air temperatures were 20.2°C. The rainy season during the study were extended from May to October with a peak during July-September. The monthly rainfall and the minimum and maximum air temperatures during the study period are presented in Figure 1a and 1b.

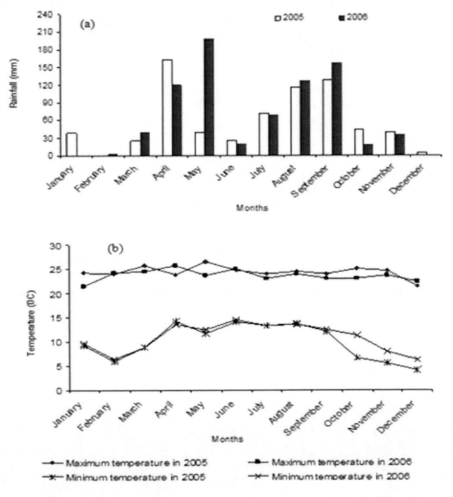

Figure 1. Monthly rainfall (1a) and maximum and minimum temperatures (1b) during the study periods at Haramaya University in the eastern part of Ethiopia.

Experimental Design and Treatments

The experiment was conducted in a randomized complete block design with 3 replications and the treatments were 2 grass species (*Chloris gayana* and *Panicum coloratum)* *and* 2 forage legumes *(Melilotus alba* and *Medicago sativa)* grown in pure stands and in mixtures using a seed rate proportions of 50:50, 67: 33, 33: 67, 75: 25 and 25: 75 of the grass and legume component, respectively. Spacing between replications and plots were 2 and 1 m, respectively. Inclusion of pure stand grasses and legumes were maintained as a control to compare the DM yield and chemical compositions with the mixtures at various seeds rates. The forage species were planted at the end of July 2005 and data on DM yield was collected for two years, until the end of 2006 growing season. Seeds of grasses and legumes were weighed, then thoroughly mixed and row planted at a spacing of 20 cm interval on 2 x 4.5 m plots. A seed rate of 10 kg ha^{-1} was used for all the pure stand grasses, forage legumes and including their mixtures. Seeding rate for each mixture was calculated according to the seed rate proportion of both grasses and legumes. Diammonium phosphate (DAP) fertiliser was applied at planting at rate of 100 kg ha^{-1} in all treatments, and 50 kg ha^{-1} nitrogen fertiliser in the form of urea was applied after establishment for pasture plots receiving only grass species according to pervious recommendations (Bogdan, 1977; IAR, 1988).

Data Collection and Analytical Procedures

Pure stand grass and legume species were harvested at about 5cm above the ground at 50 and 100% flowering, respectively while the grass-legume mixtures were harvested when at least one of the component of the mixture had reached 50% flowering based on continuous visual observations of each component in the mixture (Tessema and Baars, 2006). Immediately after harvesting the pasture in each plot, a representative sample of 500g of fresh forage sample was taken for DM yield determination by drying in an oven at 70°C for 48 h and weighing until constant weight (ILCA, 1990). The DM yield was calculated using the following formula as DMY kg ha^{-1} = (Tot FW x (DWss/FWss) x 10 (Tarawali et al., 1995) where Tot FW is the total fresh weight, DWss is dry weight of the subsample and FWss is fresh weight subsample, and the final DM yield is reported in t ha^{-1}. The relative DM yields (RY) of the components in the mixtures were calculated using the equations of De Wit (1960):

$$RY_{GL} = DMY_{GL}/DMY_{GG} \tag{1}$$

$$RY_{LG} = DMY_{LG}/DMY_{LL} \tag{2}$$

where DMY_{GG} is the DM yield of any perennial grass 'G' as a monoculture; DMY_{LL} is the DM yield of any perennial legume 'L' as a monoculture; DMY_{GL} is the dry matter yield of any perennial grass component 'G' grown in mixture with any perennial legume 'L' and DMY_{LG} is the DM yield of any perennial legume component 'L' grown in mixture with any perennial grass 'G'.

Relative total yield (RYT) was calculated according to the formula of De Wit (1960):

$$RYT_{GL} = (DMY_{GL}/DMY_{GG}) + (DMY_{LG}/DMY_{LL}) \tag{3}$$

The relative crowding coefficient (RCC) of the perennial grass-legume mixtures was calculated to determine the competitive ability of the grass and legume in the mixture to measure whether that component has produced less or more DM yield than expected in a 50:50 perennial grass-legume mixture, according to De Wit (1960):

$$RCC_{GL} = DMY_{GL}/ (DMY_{GG} - DMY_{GL}) \tag{4}$$

$$RCC_{LG} = DMY_{LG}/ (DMY_{LL} - DMY_{LG}) \tag{5}$$

For mixtures different from 50:50, RCC was calculated as:

$$RCC_{GL} = DMY_{GL} \times Zba/ (DMY_{GG} - DMY_{GL} \times Zab) \tag{6}$$

$$RCC_{LG} = DMY_{LG} \times Zab/ (DMY_{LL} - DMY_{LG} \times Zba) \tag{7}$$

where Zab is the seed rate proportion of the grass component and Zba is the seed rate proportion of the legume component in grass/legume mixture.

The dominance or aggressive ability of the perennial grasses against the perennial legumes in different seed rate mixtures was described by calculating the aggressivity index (AI) as indicated by Mc Gilchrist (1965) and Mc Gilchrist and Trenbath (1971):

$$AI_{GL} = (DMY_{GL}/DMY_{GG}) - (DMY_{LG}/DMY_{LL}) \tag{8}$$

For any replacement treatment other than 50:50, AI was calculated as:

$$AI_{GL} = (DMY_{GL}/DMY_{GG} \times Zab) - (DMY_{LG}/DMY_{LL} \times Zba) \tag{9}$$

$$AI_{LG} = (DMY_{LG}/DMY_{LL} \times Zba) - (DMY_{GL}/DMY_{GG} \times Zab) \tag{10}$$

After oven drying the fresh samples, representative thorough subsamples (250g) were taken from all the treatments for chemical analyses. Dried samples were ground to pass a 1 mm sieve and stored in airtight containers until chemical analyses. Total ash (TA) was determined by igniting at 550°C overnight, DM was determined by drying at 105°C overnight (AOAC, 1990) and nitrogen (N) by auto-analyser (Chemlab, 1984). Crude protein (CP) was calculated as N x 6.25 and organic matter (OM) as 100 - TA. Neutral detergent fibre (NDF), acid detergent fibre (ADF) and acid detergent lignin (ADL) were determined according to van Soest et al. (1991).

Hemicellulose and cellulose were calculated as NDF - ADF and ADF - ADL, respectively. All chemical analyses were done in duplicate to increase the precision of the results.

Statistical Analysis

Analysis of variance (ANOVA) was carried out using the Generalized Linear Model (GLM) procedures using SPSS statistical package (version 16.0) for DM yield, RY, RTY, RCC, AI and chemical compositions applied to a randomised complete block design. Proportional data were arcsine transformed to meet the assumption of normality and homogeneous variance prior to carrying out the ANOVA. Mean separation was done using the Duncan's Multiple Range Test (DMRT).

RESULTS

Dry Matter Yield

The DM yield of pure stand grasses were higher than pure stand forage legumes maintained as a control throughout the study. In the first year of pasture establishment, *C. gayana* and *M. alba* had a higher DM yield than all the grass-legume mixed pastures with 12 and 10.8 t ha^{-1}, respectively (Table 1). Among the pure stands, *C. gayana* had a higher DM yield than other pure stand pastures both in the first and second year of the pasture.

The DM yield of the pure stand grasses and forage legumes increased as the pasture period increased. There was a significant (P<0.05) effect on the DM yield of the grass and legume components in the mixed pasture due to seed rate proportions in the present study (Table 1). Moreover, the grass component had a higher total DM yield of the mixed pasture than the forage legume component both in the first and second year of the pasture with an overall mean DM yield of 12.2 and 8.8 t ha^{-1}, respectively. The DM yields of the grass and forage legume components increased as the pasture period increased in the present study. *C. gayana* and *P. coloratum* mixed with *M. alba* at a seed rate proportion of 50:50 and *C. gayana* mixed with *M. alba* with a seed rate proportions of 67:33 and 75: 25 provided higher mean grass DM yield component in the mixture, whereas *C. gayana* mixed with *M. alba* at a seed rate proportions of 33:67 and 50:50 gave the higher mean forage legume DM yield components of the pasture. The grass-legume mixtures had a higher mean total DM yield of pasture compared to the pure stand grasses and forage legumes maintained as a control (Table 1). The total DM yield of the grass-legume mixed pastures at different seed rate proportions was lower than the DM yield of the pure stand grasses and forage legumes in the first year of pasture establishment period. However, the total DM yield of the grass-legume mixed pastures at different seed rate proportions was higher than the DM yield of the pure stand grasses and forage legumes in the second year of the pasture. Moreover, the total DM yield of all the mixed pastures varied between the first and second years, with a progressive increase in total DM yield of the mixtures in the second year of the pasture (Table 1). 100 and 75 percent of the grass-legume mixed pastures were grater than the DM yield of pure stand legumes and grasses, respectively throughout the study period (Table 1). *C. gayana* mixed with *M. alba* at a seed rate proportion of 25:75, and *C. gayana* and *P. coloratum* mixed with *M. alba* at a seed rate proportion of 50:50 provided higher DM yield in the first year of the pasture with a DM yield of 12.9; 11.4 and 13.3 t ha^{-1}, respectively. In 2005, *C. gayana* mixed with *M. alba* at a seed rate proportion of 33:67 and 50:50 had a higher DM yield of 40.4 and

41.13 t ha^{-1}, respectively than other grass-legume mixed pastures and pure stand grass and forage legume species. Higher mean total DM yields were obtained from *C. gayana* and *P. coloratum* mixed with *M. alba* at a seed rate proportions of 50:50 with 26.7 and 25.0 t ha^{-1}, respectively, and *C. gayana* mixed with *M. alba* at a seed rate proportion of 33: 67 with 26.1 t ha^{-1} than other grass-legume mixed pastures and pure stand grasses and legumes throughout the pasture period.

Biological Potential

There was no significant (P>0.05) effect on RY of both the grass and legume components and RYT of their mixtures due to various seed rate proportions in this study.(Table 2). However, a significant (P<0.05) effect was observed on RCC and AI of both the grass and legume components and their mixtures due to different seed rate proportions (Table 2). The mean RY of both the grass and legume components were greater than unity (1.67). Moreover, the 2 years mean RYT of all grass-legume mixtures in the study period were greater than one and ranged from 1.23-2.39. *C. gayana* mixed with *M. alba and M. sativa* at a seed rate proportion of 67:33 had a higher RTY than other grass-legume mixtures in the present study (Table 2). The mean RCCs values of both the grass and legume components in the mixtures were almost equal each other and also greater than unity (Table 2). *C. gayana* mixed with *M. sativa* at a seed rate proportion of 50:50 and *P. coloratum* mixed with *M. alba* at a seed rate proportions of 50:50 and 25:75 produced a higher RCC values of the grass component, whereas *C. gayana* and *P. coloratum* mixed with *M. alba* at a seed rate proportions of 50:50 and 25:75, respectively produced higher RCC values of the legume component compared to other grass-legume combinations. More than 75% of the grass/legume mixtures produced mean AI values of less than unity and closer to zero (range: -0.06 to +0.88) and the mean AI value of both the grass and legume components was low (0.29) (Table 2).

Chemical Composition

A significant (P<0.05) difference on CP, NDF and hemicellulose contents was observed between pure stand grasses and legumes, and their mixtures in the present study. However, there was no significant effect on DM, OM, TA, ADF, ADL, cellulose contents between pure stand grasses and legumes, and their mixtures (Table 3). All pure stand legumes and their mixture with grasses had a higher CP contents than pure stand grasses. On the contrary, pure stand grasses had a higher NDF contents compared to pure legumes and their mixtures. The CP contents of all pure stand grasses and legumes as well as their mixtures at various seed rate propitiations was greater than 18%, ranged from 17.92%-23.6%. Moreover, the NDF contents of all pure stand grasses and legumes as well as their mixtures at various seed rate propitiations were lower than 60% and ranged from 44.0%-58.8%. The CP and NDF contents of the grass-legume mixed pastures indicate that all the mixtures are categorized under good quality forages. All pure stand forage legumes and *C. gayana* mixed *Medicago* sativa at a seed rate proportion of 50:50 and *P. coloratum* and *C. gayana* mixed with *M. sativa* both at seed rate proportions of 25:75 provided a higher CP contents compared to pure stand grasses and other mixtures in the present study (Table 3).

Table 1. Dry matter yield (t ha^{-1}) of *Chloris gayana* and *Panicum coloratum* mixed with *Melilotus alba* and *Medicago sativa* at various seed rate proportions at Haramaya University of eastern Ethiopia

	Grass	Legume	Total	Grass	Legume	Total	Grass	Legume	Total
Chloris gayana (R)	12.0^a	-	12.0^{abc}	24.0^{ab}	-	24.0^{cdef}	18.0^a	-	18.0^{bcdef}
Panicum coloratum (P)	7.3^{bc}	-	9.8^{abcdefgh}	16.5^{bcdefg}	-	20.1^{def}	14.9^{ab}	-	14.9^{ef}
Melilotus alba (M)	-	10.8^a	10.8^{abcdefgh}	-	18.3^{ab}	18.3^{ef}	-	14.6^a	14.6^{ef}
Medicago sativa (A)	-	8.6^{ab}	8.6^{abcdefgh}	-	14.8^{abcde}	14.8^f	-	11.7^{abc}	12.6^f
R: M (50:50)	7.4^{bc}	5.5^{bcd}	12.9^{ab}	24.3^a	16.1^{abcd}	40.4^a	15.9^{ab}	10.8^{bcd}	26.7^a
R: A (50:50)	3.5^{cde}	4.1^{cde}	7.6^{fgh}	23.5^{abc}	12.4^{bcdef}	35.9^{abc}	13.5^{abcd}	8.3^{cdefg}	21.8^{abcde}
P: M (50:50)	8.3^{ab}	3.1^{cde}	11.4^{abcde}	21.7^{abcde}	16.9^{abc}	38.6^{ab}	15.0^{ab}	10.0^{bcde}	25.0^{abc}
P: A (50:50)	5.5^{bcde}	4.6^{cde}	10.1^{abcdefgh}	16.0^{cdefg}	10.3^{def}	26.3^{bcdef}	10.7^{bcde}	7.5^{defg}	18.2^{bcdef}
R: M (67:33)	5.6^{bcde}	5.5^{bcd}	11.0^{abcdefg}	23.1^{abcd}	12.6^{bcdef}	35.7^{abc}	14.3^{abcd}	9.0^{bcdefg}	23.4^{abcd}
R: M (33:67)	5.7^{bcde}	5.4^{bcd}	11.1^{abcdefg}	21.5^{abcde}	19.6^a	41.1^a	13.6^{abcd}	12.5^{ab}	26.1^{ab}
P: M (67:33)	5.6^{bcde}	2.6^{cde}	8.2^{cdefgh}	15.6^{defg}	11.7^{cdef}	27.4^{bcdef}	10.6^{bcde}	7.2^{efg}	17.8^{cdef}
P: M (33:67)	6.5^{bcde}	4.4^{cde}	10.8^{abcdefg}	16.0^{cdefg}	16.6^{abc}	32.6^{abcd}	11.2^{abcde}	10.5^{bcde}	21.7^{abcde}
P: A (25:75)	3.6^{cde}	4.7^{cde}	8.3^{cdefgh}	9.6^g	12.5^{bcdef}	22.1^{def}	6.6^e	8.6^{cdefg}	15.2^{def}
P: A (75:25)	6.1^{bcde}	3.5^{cde}	9.7^{abcdefgh}	17.5^{abcdef}	7.8^f	25.3^{cdef}	11.8^{abcde}	5.7	17.5^{cdef}
P: M (25:75)	3.2^{de}	4.2^{cde}	7.4^{gh}	12.1^{fg}	14.8^{abcde}	26.9^{bcdef}	7.7^{cde}	9.5^{bcdef}	17.2^{cdef}
P: M (75:25)	7.8^e	1.8^e	9.6^{abcdefgh}	20.2^{abcde}	9.7^{ef}	29.8^{abcde}	14.0^{abcd}	5.7	19.7^{abcdef}
R: M (25:75)	7.2^{bcd}	6.0^{bc}	13.3^a	15.3^{efg}	14.8^{abcde}	30.1^{abcde}	11.3^{abcde}	10.4^{bcde}	21.7^{abcde}
R: M (75:25)	7.0^{bcd}	2.2^{de}	9.3^{bcdefgh}	21.8^{abcde}	10.6^{def}	32.4^{abcd}	14.4^{abc}	6.4^{fg}	20.8^{abcdef}
R: A (25:75)	2.7^e	4.1^{cde}	6.8^h	12.2^{fg}	11.6^{cdef}	23.8^{cdef}	7.4^{de}	7.9^{defg}	15.3^{def}
R: A (75:25)	7.6^{bc}	2.7^{cde}	7.8^{efgh}	19.7^{abcdef}	8.4^f	24.6^{cdef}	10.7^{bcde}	5.6^g	16.2^{def}
R: A (33:67)	6.5^{bcde}	5.3^{bcd}	11.7^{abcd}	16.5^{bcdefg}	13.4^{bcdef}	29.9^{abcde}	11.5^{abcde}	0.3^{bcdef}	20.8^{abcde}
R: A (67:33)	5.8^{bcde}	4.8^{cde}	10.6^{abcdefg}	21.1^{abcde}	9.3^{ef}	30.4^{abcde}	13.5^{abcde}	7.0^{efg}	20.5^{abcde}
P: A (33:67)	5.8^{bcde}	4.0^{cde}	9.8^{abcdefgh}	18.2^{abcdef}	11.0^{cdef}	29.2^{abcde}	12.0^{abcde}	7.5^{defg}	19.5^{abcdef}
P: A (67:33)	6.1^{bcde}	5.3^{bcd}	11.3^{abcdef}	14.6^{efg}	10.3^{def}	24.9^{cdef}	10.3^{bcde}	7.8^{defg}	18.1^{bcdef}
Mean	6.2	4.7	10.1	18.2	12.9	28.5	12.2	8.8	19.3
SEM	1.42	0.90	1.58	3.24	1.55	3.38	1.74	0.90	1.86

Within columns, means followed by the same letter are not significantly different at $P = 0.05$.

Table 2. Effect of various seed rate proportions on relative yield (RY), relative total yield (RYT), relative crowding coefficients (RCC), and aggressivity index (AI) of the grass and herbaceous legumes grown in mixtures at Haramaya University, eastern Ethiopia

	RY[1] (grass)	RY (legume)	RYT[2]	RCC[3] (grass)	RCC (legume)	AI[4] (grass)	AI (legume)
R: M[¶] (50: 50)	0.70	0.53	1.23	3.46	2.58	0.06	-0.06
R: A (50: 50)	0.64	0.64	1.28	1.74	1.15	-0.07	0.07
P: M (50: 50)	1.26	0.64	1.90	6.84	2.63	0.52	-0.52
P: A (50: 50)	0.86	0.95	1.81	2.03	1.16	0.50	-0.50
R: M (33: 67)	1.23	0.71	1.95	0.73	1.86	-0.59	0.59
R: M (67: 33)	1.30	1.08	2.39	1.99	3.01	0.41	-0.41
P: M (67: 33)	0.93	0.56	1.49	1.79	1.79	-0.09	0.09
P: M (33: 67)	1.00	0.83	1.82	0.44	1.59	1.56	-1.56
P: A (25: 75)	0.63	1.09	1.72	1.81	1.96	1.28	-1.28
P: A (75: 25)	0.77	0.64	1.41	0.46	1.37	-0.42	0.42
P: M (25: 75)	0.70	0.75	1.44	3.54	5.20	1.33	-1.33
P: M (75: 25)	1.08	0.35	1.43	0.68	1.53	0.09	-0.09
R: M (25: 75)	0.64	0.84	1.48	1.25	1.45	1.27	-1.27
R: M (75: 25)	0.92	0.60	1.52	1.09	1.89	-0.59	0.59
R: A (25: 75)	0.42	0.96	1.38	0.90	0.67	0.34	-0.34
R: A (75: 25)	0.73	0.76	1.49	0.22	1.42	-0.88	0.88
R: A (33: 67)	0.87	1.08	1.95	1.75	1.44	0.63	-0.63
R: A (67: 33)	0.67	1.45	2.11	0.52	1.44	-0.44	0.44
P: A (33: 67)	1.29	0.70	1.99	1.29	0.61	1.52	-1.52
P: A (67: 33)	0.78	0.78	1.56	1.19	1.76	-0.55	0.55
Mean	0.87	0.80	1.67	1.70	1.71	0.29	-0.29
SEM	0.25	0.20	0.35	1.97	0.84	0.30	0.30
Probability	NS	NS	NS	<0.001	<0.001	<0.001	<0.001

[¶] A = Medicago sativa; M = Melilotus alba; P = Panicum coloratum; R = Chloris gayana

[1] Relative yield = Yield when grown in a mixture relative to yield as pure stand.

[2] Relative yield of grass plus relative yield of legume.

[3] Relative crowding coefficient = Yield when grown in a mixture as a proportion of (yield in pure stand less yield in mixture).

[4] Aggressivity index = (Actual yield of component/Expected yield of component) minus (Actual yield of other component/Expected yield of other component).

Table 3. Chemical composition (% in DM basis) of *Chloris gayana* and *Panicum coloratum* mixed with *Medicago sativa* and *Melilotus alba* at various seed rate proportions at Haramaya University of eastern Ethiopia

	Chemical composition (% DM basis)[£]								
	CP	DM	OM	TA	NDF	ADF	ADL	Cell	HC
Chloris gayana (R)	18.5[efg]	93.0	86.6	13.4	55.7[ab2]	33.8	5.9	28.0	21.9[ab]
Panicum coloratum (P)	17.9[g]	93.4	87.2	12.8	58.8[a]	34.4	5.3	29.2	24.4[a]
Melilotus alba (M)	23.1[ab]	93.1	88.2	11.8	46.5[efg]	47.4	6.9	40.5	11.1[f]
Medicago sativa (A)	23.6[a]	93.1	87.4	12.6	51.1[bcdefg]	48.6	6.9	41.6	14.9[cdef]
R: M (50:50)	18.6[efg]	93.5	89.1	10.9	48.9[bcdefg]	35.6	6.4	29.2	13.27[ef]
R: A (50:50)	22.2[abc]	93.0	88.0	12.0	46.9[defg]	31.1	6.3	24.8	15.8[bcdef]
P: M (50:50)	20.37[cdefg]	93.3	87.7	12.3	54.9[abc]	34.7	6.2	28.5	20.2[abcd]
P: A (50:50)	21.0[abcde]	93.6	88.3	11.7	46.2[efg]	34.7	6.8	27.9	11.6[ef]
R: M (67:33)	19.8[cdefg]	93.0	88.2	11.9	49.1[bcdefg]	35.0	5.8	29.3	14.1[cdef]
R: M (33:67)	18.3[efg]	91.9	88.0	12.0	47.9[cdefg]	37.0	6.3	30.7	14.3[cdef]
P: M (67:33)	19.3[defg]	93.1	88.2	11.8	51.2[bcdef]	34.9	6.7	28.5	16.3[bcdef]
P: M (33:67)	18.3[efg]	93.5	87.4	12.7	52.9[abcde]	35.7	6.4	29.3	17.2[bcdef]
P: A (25:75)	21.6[abcd]	92.9	87.7	12.3	44.0[g]	33.3	6.2	27.1	10.7[f]
P: A (75:25)	20.0[cdefg]	93.0	87.0	13.0	52.1[abcdef]	34.9	6.5	28.4	17.1[bcdef]
P: M (25:75)	19.9[cdefg]	93.1	87.4	12.6	45.9[efg]	34.4	6.0	28.5	11.5[f]
P: M (75:25)	19.0[defg]	93.0	89.1	12.6	54.1[abcd]	33.5	5.5	28.1	20.5[abc]
R: M (25:75)	18.6[efg]	92.8	86.3	10.9	48.8[bcdefg]	34.6	6.0	28.0	14.2[cdef]
R: M (75:25)	18.2[fg]	93.4	87.9	13.7	52.4[abcdef]	34.1	5.4	28.8	18.3[abcde]
R: A (25:75)	20.9[abcdefg]	92.6	87.7	12.1	45.6[fg]	32.5	6.0	26.5	13.16[ef]
R: A (75:25)	19.9[cdefg]	93.1	87.9	12.3	49.4[bcdefg]	33.0	6.6	26.4	16.4[bcdef]
R: A (33:67)	19.6[cdefg]	93.2	87.4	12.1	49.0[bcdefg]	33.4	6.4	27.0	15.6[bcdef]
R: A (67:33)	20.69[bcdefg]	93.5	88.4	11.6	45.9[bcdefg]	32.2	6.4	25.8	13.7[def]
P: A (33:67)	20.2[cdefg]	93.4	87.6	12.4	47.5[efg]	31.6	6.4	25.2	15.9[bcdef]
P: A (67:33)	19.4[defg]	92.9	87.2	12.8	50.5[bcdefg]	33.1	6.4	26.7	17.4[bcdef]
Mean	19.9	93.1	87.7	12.3	48.6	35.1	6.2	28.9	13.4
SEM	0.97	0.37	0.64	0.64	2.52	3.74	0.43	3.82	4.66

[£]ADF = Acid detergent fibre; ADL = Acid detergent lignin; CP = Crude protein DM = Dry matter; OM = Organic matter; TA= Total Ash; NDF = neutral detergent fiber; Cellu = Cellulose; HC = Hemicelluloses.
Within columns, means followed by the same letter are not significantly different at $P = 0.05$.

DISCUSSION

Dry Matter Yield

In the current study, *C. gayana* and *P. coloratum* mixed with *M. alba* at a seed rate proportions of 50:50, and *C. gayana* mixed with *M. alba* at a seed rate proportion of 33: 67 provided than a higher DM yield productions than other grass-legume mixed pastures and pure stand grasses and legumes throughout the pasture period. Moreover, 100 and 75% of the grass-legume mixed pasture was higher than the total DM yield of pure stand grass and herbaceous forage legume species, respectively, indicating the contribution of various seed rate proportions of both pure stand grass and forage legume species on the total DM yield of their grass-legume mixtures. The results of the total DM yield of the grass-legume mixed pastures in the present study are in agreement with previous reports (Tessema, 1996; Diriba, 2002; Tessema and Baars, 2006). However, according to Lemma et al. (1991) pure stand *C. gayana* produced a higher DM yield than a *C. gayana*-forage legume mixed pasture in the second and third year of establishment in the western parts of Ethiopia.

The total DM yields of all grass-legume mixed pastures varied between the first and second years, and total DM yield production increased as the pasture sward increases. This might be due to the fact that all the grass and forage legume species as pure stands and in the mixtures grew well and vigour's from the time of establishment to flowering because grasses and legumes usually produce more tillers and branches, respectively that could contribute to the higher total DM yield for the grass-legume mixed pastures. In addition, forage legumes had fast regrowth after each harvest than grass species that could contribute to a higher total DM yield productions of the grass-legume mixtures throughout the study (personal observation).

Moreover, as the growth pattern of forage legumes and grasses is variable over drier and wetter parts of the year, selection of the correct grass-legume combination could make sustained animal feed available year round under smallholder farming conditions in Ethiopia. This indicates that higher total DM yields of grass-legume combinations could be obtained as the pasture sward advances. Previous studies elsewhere in Ethiopia and other tropical areas of world have indicated that grass and legume mixed pastures could significantly contribute to a higher total herbage yield (Morrison, 1984; Daniel, 1996; Lemma et al., 1993, Tessema 1996; Tessema and Baars, 2006). The DM yield from natural pasture over sown by *Stylosanthes guianensis* was found to be promising in subtropical areas of Ethiopia (Lemma et al., 1991, 1993; Diriba, 2002). Principally, when a pure grass pasture is grown tie up of N in below ground parts, it eventually suffers yield losses through N depletion. Conversely pure legume pasture fixes N in excess of its requirement that attract insects invading non-legume weeds or grasses (Lemma et al., 1991).

Biological Potential

The 2 years mean RYT of all grass-legume mixtures in the experimental periods were greater than one (range: 1.23-2.39) reflecting that there was a yield advantage of 23-139 percent over the pure stands of grasses and legumes. This may probably suggest the occurrence of biological nitrogen fixation in the root nodules of the legumes and its transfer from the legume component to the grasses that might have supported the growth of grasses and their mixture in the pasture sward. This would also imply that the grass-legume mixtures were, at least, partly complementary in resource use. This may happen when the growth periods of the mixtures are partly overlapping or when they are using plant growth resources from varying soil depths. The results of the RY and RTY in the current study were in agreement with Daniel (1996) and Diriba (2002) for the *Chloris-Medicago* and *Panicum-Stylosanthes* mixtures, respectively. Tessema and Baars (2006) also reported a RYT values to have ranged from 1.29 to 1.48 in a 50:50% perennial grass-legume mixtures in the north-western part of Ethiopia. This situation could be attributed to the efficient utilization of plant growth factors by species in the mixture due to either temporal or spatial differences of their demands.

Both grasses and legumes in the mixture produced RCCs in excess of unity indicating that all produced a higher yield in the mixture than expected in pure stands at different seed rates. Moreover, the mean RCCs for both the grass and legume components were almost equal indicating that they perform similarly in DM production in the mixture. This may indicate that the grasses were highly benefited highly from the legume components, which could contributed to increased DM productions of the mixtures. Forage legumes benefits in terms of increased herbage and animal production in smallholder agricultural production systems by their ability to fix atmospheric nitrogen (Thomas, 1995). Morrison (1984) also suggested that legumes such as *Medicago* and clovers increase the yield of the grasses when grown in combination with them. More than 75% of the grass/legume mixtures produced mean AI values of less than unity and closer to zero (range: -0.06 to +0.88) and the mean AI value of both the grass and legume components was low (+0.29 and -0.29, respectively) indicating that there was small dominance in both components and they were almost equally competitive and both contributed more in DM production of the entire mixture throughout the study. The AI values of the present study is similar with the result of Tessema and Baars (2006) who reported that the AI values for *Chloris* mixed with *Desmodium* and *Chloris* mixed with *Medicago* were +0.07 and +0.23, respectively.

Chemical Composition

As expected, all pure stand legumes and their mixtures with grasses had higher CP values compared to pure stand grasses in the present study. The CP level required for maintenance and production of cattle is 150 g kg^{-1} DM (Norton, 1982; Van Soest, 1982) indicating that all the pure grasses and legumes, and their mixtures are above (range 179.2-236.4 g kg^{-1} DM) the recommendation. Pure grasses had higher NDF, ADF, cellulose and hemicellulose values compared to pure legumes and their mixtures in this study. The leaf stem ration of the grass and legume components in the mixture was greater than unity in the present study (Tessema, 2008) and this might have contributed to the higher CP and lower fibre fractions in all grass/legume mixtures compared to pure stand grasses. The threshold level of NDF in tropical grass beyond which DM intake of cattle is affected is 600 g kg^{-1} DM (Meissner et al., 1991); however, all the pure grasses and legumes, and their mixtures have lower than this value (range: 440-588 g kg^{-1} DM). In the present study, the hemicellulose (range 107.0-240.0 g kg^{-1}

DM) contents of all the pure grasses and legumes, and their mixtures were lower than those of most tropical grasses, 354 g kg^{-1} DM (Moore and Hatfield, 1994).

Accordingly a combination of grasses and legumes is therefore usually far preferable to any mixture of grasses alone, without legumes so that they may utilize fixed N that improves the feeding value of the sward (Morrison, 1984; Larbi et al., 1995; Tesssema, 1996). This showed that the forage legume increases the digestibility and the palatability of the grass component and the mixed pasture sward as a whole (Tessem and Baars, 2004, 2006). Grass and legume mixtures significantly contribute to the quality of the pasture sward as reported by many authors (Daniel 1990; Lemma et al 1993; Diriba 2002; Tessema and Baars, 2004, 2006). The animal production from natural pasture over sown by *Stylosanthes guianensis* was found to be promising in subtropical areas of Ethiopia (Lemma et al., 1993). The other advantage of perennial legumes in a mixed pasture is their higher protein value at any given time than the accompanying grass species (Lemma et al., 1991). Grass-legume mixtures improved forage quality due to the high nutrient concentration in the legumes (Larbi et al., 1995; Getnet, 1999). Also, legumes are important to increase the amount of crude protein, extend the grazing periods of the dry season and provide nitrogen for the companion grasses (Crowder and Chedda,m 1982; Daniel, 1996, Tessema and Baars, 2006). Crowder and Chedda (1982) indicated that *Chloris and Molasses* grasses have high initial dry matter digestibility (DMD) (70-85%) but this sharply declines with the advancement of maturity. Similarly, Daniel (1996) and Larbi et al. (1995) reported a higher CP and in vitro dry matter digestibility of *C. gayana* mixed with *M. sativa* and a decline of CP and IVDM after maturity in the central highlands of Ethiopia. However, the reduction of DMD with the advancement of maturity could be improved by growing grasses with perennial forage legumes such as *Desmodium* and *Medicago* (Larbi et al., 1995). This might be due to the fact that the grass-legume mixtures increase CP through atmospheric nitrogen fixation and then reduces the fiber fractions of the pasture, which are indicators of good quality forages (Mannteje, 1984; Minson, 1990) and thereby this increases DMD. This showed that the forage legumes increase the quality of the grass component and the mixed pasture sward as a whole (Tessema and Baars, 2004, 2006). The combinations of grass and forage legume species are therefore far preferable to any pure stand grass swards because the forage legume species in the mixture can fix N and the grasses may utilize the fixed N that improves the feeding value of the grass-legume sward (Morrison, 1984). This may probably suggest the occurrence of biological nitrogen fixation in the root nodules of the legumes and its transfer from the legume component to the grasses that might have supported the growth of grasses and their mixture in the pasture sward. This also implies that the grass-legume mixtures are, at least, partly complementary in resource use and this may usually happen when the growth periods of the mixtures are partly overlapping or when they are using plant growth resources from varying soil depths (Morrison, 1984; Larbi et al., 1995).

CONCLUSION

The results of the present study revealed that a higher total DM yield of grass-legume mixed pasture than pure stand grass and legume species. *C. gayana* and *P. coloratum* mixed with *M. alba* at a seed rate proportion of 50:50 and *C. gayana* mixed with *M. alba* at a seed rate proportion of 33:67, respectively had a higher total DM yield than other mixtures and pure stand grasses and legumes in the present study. Moreover, the present study confirmed that both forage legumes and their mixtures with perennial grass species had a higher CP and a lower fiber contents compared to pure stand grass species. As a conclusion, to alleviate the feed shortage situation, therefore it is recommended that *C. gayana* mixed with *M. alba* at a seed rate proportions of 50: 50 and 33:67 can be introduced under smallholder crop-livestock farming system conditions in the eastern parts of Ethiopia. In addition, further studies on animal performances should be needed using feeding experiments and under grazing conditions to see how these mixtures perform in practical situations.

ACKNOWLEDGMENTS

The author would like to acknowledge the Haramya University (HU) of Ethiopia and the Ethiopian Agricultural Research Institute, for financing the research and all staff members of the Animal Nutrition Laboratory of HU for their assistance during chemical analyses.

REFERENCES

AOAC (Association of Official Analytical Chemists) (1990) Official methods of analysis, fifth edition, AOAC INC: Arlington, Virginia, USA.

Berhan, T (2005) Biological potential and economic viability of grass (Panicum coloratum) and legume (Staylosanthes guinensis) association under variable seed rates in eastern highlands of Ethiopia. *Tropical Science*, 45, 83-85.

Berhan, T. (2006) Biological potential and economic viability of grass (Chloris gayana) and legume (Stylosanthes guianensis) associations under variable seed rates in the eastern highlands of Ethiopia. *Tropical Science*, 46, 61-63.

Bogdan, A. (1977) Tropical Pasture and Fodder plants (Grasses and legumes). London: Longman Inc.

Chemlab (1984) Continuous flow analysis system 40. Method sheet No. CW2-008-17 (Ammonia (0-1 and 0-50 PPM. N), Horn Church, Essex, UK: Chemlab Instruments Ltd.

Crowder, LV., Chheda, H.R. (1982) Tropical grassland husbandry, New York: Longman Inc

Daniel, K. (1996) Effects of nitrogen application and stage of development on yield and nutritional value of Rhodes Grass (Chloris gayana). *Ethiopian Journal of Agricultural Science*, 15, 86-101.

De Wit, C.T. (1960) On competition. *Verslag Landbouwkundig Onderzoek*, 66, 1-82.

Diriba, G. (2002) *Panicum coloratum* and *Stylosanthes guianensis* mixed pasture under varying relative seed proportion of the component species: yield dynamics and inter-component interaction during the year of establishment. In: Livestock in food security-roles and contributions, proceedings of the 9th annual conference of the Ethiopian Society of Animal Production (ESAP) held in Addis Ababa, Ethiopia, August 30-31, 2001, pp. 285-293.

Getnet, A. (1999) Feed resource assessment and evaluation of forage yield, quality and intake of oats and vetches grown in pure stands and in mixtures in the high lands of Ethiopia. MSc Thesis, Uppsala, Sweden: Swedish University of Agricultural Sciences.

IAR (1988) Handbook on Forage and Pasture Crops for Animal Feeding. Addis Ababa, Ethiopia: Institute of Agricultural Research

ILCA (International Livestock Center for Africa), (1990) Livestock research manual. Addis Ababa, Ethiopia: ILCA.

Larbi, A., Ochang, L. and Addie, A. (1995) Dry matter production pf thirteen tropical legumes in association with Rhodes grass (Chloris gayana) on acid soil in Ethiopia. *Tropical Grasslands,* 29, 88-91.

Lemma, G., Alemu, T. and Liyusew, A. (1991) Evaluation of different grass-legume mixtures in the mid sub humid zones of Ethiopia. Proceedings of the 4th National Livestock Improvement Conference, Addis Ababa, November 1991, pp: 13-15.

Lemma, G., Alemu, T. and Abubeker, H. (1993) Botanical composition, improvement interventions and cattle weight gains of natural pastures of western Ethiopia. Proceedings of the XVII International Grassland Congress, New Zealand and Australia, 1993, pp. 309-311.

Mc Gilchrist, C.A. (1965) Analysis of competition experiments. *Biometrics,* 21, 975-985.

Mc Gilchrist, C.A. Trenbath, B.R. (1971) A revised analysis of plant competition experiments. *Biometrics,* 27, 659-677.

Mannetje, L.T. (1984) Nutritive values of tropical and subtropical pastures, with special reference to protein and energy deficiency in relation to animal production. In: McGilchrist CA and Mackie (Eds.) Herbivore Nutrition in Subtropics and Tropics. Gaighance, South Africa: Science, Press Pty Ltd.

Meisser, H.H., Koster, H.H., Nieuwoudt, S.H. Coetze, R.J. (1991) Effects of energy supplementation on intake and digestion of early and mid-season ryegrass and Panicum/Smuts finger hay, and on in sacco disappearance of various forage species. *South African Journal of Animal Science,* 21, 33-42.

Mekoya, A.K. (2008) Multipurpose fodder trees in Ethiopia: Farmers' perception, constraints to adoption and effects of long-term supplementation on sheep performance. PhD thesis, Wageningen, The Netherlands: Wageningen University.

Minson, D.J. (1990) The chemical composition and nutritive values of tropical grasses. In: Skerman, P.J., Riveros, F. (Eds.) Tropical grasses, Rome, Italy: FAO.

Moore, K.J. and Hatfield, R.D. (1994) Carbohydrates and forage quality. In. Fahey, G.C., Collins, M.D., Mertens, R. and Moser, E. (Eds.) Forage Quality, Evaluation and Utilisation, Madison, USA: American Society of Agronomy, Crop Science of America and Soil Science Society of America.

Morrison, F.B. (1984) Feeds and Feeding: A handbook for the students and stockman. New Delhi, India: CBS publishers and distributors.

Norton, B.W. (1982) Differences between species in forage quality. Proceedings of the international symposium held on 24-28 September 1981. St. Lucia, Queensland, Australia, pp. 76-84.

Prasad, N.K., Bhagat, R.K. Singh, A.P. and Sing,R.S. (1991) Intercropping of Deenanath grass (*P. pedicellatum*) with cowpea (*V. unguiculata*) for forage production. *Indian Journal of Agricultural Sciences,* 60, 15-18.

Tamire, H. (1982) Summary of results of soil research programme in Hararghe highlands, eastern Ethiopia. Addis Ababa University, Ethiopia: Department of Plant Sciences, College of Agriculture.

Tarawali, S.A., Tarawali, G., Larbi, A. and Hanson, J. (1995) Methods for the evaluation of legumes, grasses and fodder trees for use as livestock feed. Nairobi, Kenya: International Livestock Research Institute (ILRI).

Tessema, Z. (1996) Forage yield performance of different perennial grass-legume mixtures in Northwestern Ethiopia. Proceedings of the 4th national conference of the Ethiopian Society of Animal Production (ESAP) held in Addis Ababa, pp: 203-207.

Tessema, Z. and Baars, R.M.T. (2004) Chemical composition, *in vitro* dry matter digestibility and ruminal degradation of Napier grass (*Pennisetum purpureum* (L.) Schumach.) mixed with different levels of *Sesbania sesban* (L.) Merr.). Animal Feed Science and Technology, 117, 29-41.

Tessema, Z. and Baars, R.M.T. (2006) Chemical Composition and dry matter production and yield dynamics of tropical grasses mixed with perennial forage legumes. *Tropical Grasslands,* 40, 150-156.

Tessema, Z. (2008) Effect of plant density on morphological characteristics, dry matter production and chemical composition of Napier grass (*Pennisetum purpureum* (L.) Schumach). *East African Journal of Sciences,* 2, 55-61.

Van Soest, P.J. (1982) Nutritional ecology of ruminants. Oregon, USA: O and B Books.

Van Soest, P.J. Roberston, J.B. and Lewis, B.A. (1991) Methodes of dietary fibre and non-starch polysaccharides in relation to animal nutrition. *Journal of Dairy Science,* 74, 3582-3592.

Yeshitila, A. Tessema, Z. and Azage, T. (2008a) Availability of livestock feed resources in Alaba Woreda, Southern Ethiopia. 2008. In: Proceedings of the 16th annual conference of the Ethiopian Society of Animal Production (ESAP), October 8-9, 2008, Addis Ababa, Ethiopia, pp. 21-32. 26.

Yeshitila, A. Tessema, Z. and Azage, T. (2008b) Utilization of feeds, livestock unit versus dry matter requirement in Alaba, Southern Ethiopia. Proceedings of the 16th Annual Conference of the Ethiopian Society of Animal Production (ESAP), October 8-9, 2008, Addis Ababa, Ethiopia, pp. 33-38.

In: Agricultural Research Updates. Volume 5
Editors: P. Gorawala and S. Mandhatri

ISBN: 978-1-62618-723-8
© 2013 Nova Science Publishers, Inc.

Chapter 5

GENETIC VARIATION OF GLOBULIN COMPOSITION IN SOYBEAN SEEDS

Kyuya Harada[1], Masaki Hayashi[1] and Yasutaka Tsubokura[2]

[1] National Institute of Agrobiological Sciences, Tsukuba, Ibaraki, Japan
[2] Sow Brand Seed Co., Ltd., Chiba-city, Chiba, Japan

ABSTRACT

About seventy per cent of total protein in soybean seeds consists of two major storage proteins, glycinin (11S globulin) and β-conglycinin (7S globulin). Glycinin is composed of five subunits, $A_{1a}B_{1b}$, $A_{1b}B_2$, A_2B_{1a}, (group I), A_3B_4 and $A_4A_5B_3$ (group II). Beta-conglycinin is composed of three subunits, α, α' and β.

A mutant line lacking β-conglycinin was obtained by γ-irradiation. The deficiency was controlled by a single recessive allele, *cgdef* (*β-conglycinin-deficient*). The result of Southern and northern blot analyses indicated that the deficiency is not caused by a lack of, or structural defects in the β-conglycinin subunit genes, but rather arise at the mRNA level. It was assumed that the mutant gene encodes a common factor that regulates expression β-conglycinin subunit genes. Another β-conglycinin deficient mutant QT2 was identified from a wild soybean. The phenotype was found to be contolled by a single dominant gene *Scg-1* (*suppressor of β-conglycinin*). The physical map of the *Scg-1* region covered by lambda phage genomic clones revealed that the two α subunit genes were arranged in a tail-to-tail orientation, and the genes were separated by 197 bp in *Scg-1* compared to 3.3 kb in the normal allele. Moreover, small RNA was detected in immature seeds of the mutants by northern blot analysis using an RNA probe of the α subunit. These results strongly suggest that β-conglycinin deficiency in QT2 is controlled by post-transcriptional gene silencing through the inverted repeat of the α subunits.

A glycinin deficient line was developed by combining $A_4A_5B_3$ subunit deficiency, an A_3B_4 deficient line in wild soybean and a group I deficient line produced by γ-irradiation. Each deficiency was controlled by a single recessive gene.

A line lacking glycinin and β-conglycinin was generated by combination of a glycinin deficient line and the *Scg-1* gene. This line grows normally and contains large amount of free amino acids, especially arginine.

In this chapter, ↵ we will describe the characteristics of these lines and discuss their uses and significance in soybean breeding.

INTRODUCTION

Soybean seeds contain between 35 and 45% protein on a dry weight basis, of which about 70% consists of the two major storage proteins, glycinin (11S globulin) and β-conglycinin (7S globulin). Glycinin is a simple hexameric protein with a molecular weight of about 350,000 consisting of six subunits, each having an acidic (A) and a basic (B) polypeptide component linked by a single disulfide bond. Five major glycinin subunits, $A_{1a}B_{1b}$, $A_{1b}B_2$, A_2B_{1a} (group I), A_3B_4 and $A_4A_5B_3$ (group II), have been identified. The $A_{1a}B_{1b}$ and A_2B_{1a} subunits are encoded by *Gy1* and *Gy2* respectively which are closely located on chromosome 3. The $A_{1b}B_2$, $A_4A_5B_3$ and A_3B_4 subunits are encoded by *Gy3* (chromosome 19), *Gy4* (chromosome 10) and *Gy5* (chromosome 13) respectively (Nielsen et al. 1997, Beilinson et al. 2002, Li and Zhang 2011).

Beta-conglycinin is a trimeric glycoprotein, and is composed of three subunits, α, α' and β in varing propotions. *CG-1* on chromosome 10 encodes the α' subunit (Harada et al. 1989), *CG-2* and *CG-3* on chromosome 20 encode the α subunits (Yoshino et al. 2002, Li and Zhang 2011, Tsubokura et al. 2012), and *CG-4* and another gene on chromosome 20 encode the β subunit (Harada et al. 1989, Li and Zhang 2011, Tsubokura et al. 2012).

Glycinin contains more sulfur-containing amino acids than β-conglycinin (Fukushima 1968). There are different characteristics in protein food processing between glycinin and β-conglycinin. In addition, the subunit compositions of the glycinin markedly effect the hardness, turbidity, and rates of gelation of the glycinin-gels (Fukushima 1991). The tofu-gel from crude glycinin was remarkably harder than that from crude β-conglycinin in the presence of calcium sulfate (Saio et al. 1969) due to the higher number of SS bonds in glicinin tofu-gel than in β-conglycinin gel (Saio et al. 1971). Tofu-gel from β-conglycinin had weaker strength and was softer than tofu-gel from glycinin in the presence of glucono-delta-lactone (Kohyama and Nishinari 1993). Utsumi and Kinsella (1985) reported that β-conglycinin-gel was harder than glycinin-gel when solutions were heated at 80°C for 30 min. Soymilk from the low-β-conglycinin soybean coagulated immediately after addition of 0.3% MgCl2 was added, while soymilk from the low-glycinin soybean remained liquid even at 1°C(Yagasaki et al. 2000). The A_3B_4 glycinin subunit played an important role in increasing gel hardness, whereas the $A_4A_5B_3$ glycinin subunit was closely related to the gel formation rate in the heat-induced gels (Nakamura et al. 1984, 1985). The viscosity of the soymilk with $A_4A_5B_3$ subunit, after magnesium chloride addition, was higher than that of the soymilk without $A_4A_5B_3$ subunit (Murasawa et al. 1991). The tofu-gel prepared from the soymilk without $A_4A_5B_3$ subunit was harder than that prepared from soymilk with $A_4A_5B_3$ subunit, regardless of coagulant types and their concentrations (Murasawa et al. 1991). Similar results were obtained by using tofu-gel in the presence of glucono-delta-lactone (Nishinari et al. 1991).

A peptide with the sequence VLIVP from the glycinin was found to have angiotensin I-converting enzyme (ACE) inhibitory property (Mallikarjun Gouda et al. 2006). ACE inhibitory peptides isolated from the β-conglycinin hydrolysate were identified as LAIPVNKP and LPHF, and those from glycinin hydrolysate as SPYP and WL (Kuba et al. 2005). The serum cholesterol lowering effect depending upon the dietary level in rats was observed in the crude and highly purified glycinin (Urade and Kohno 2011). The α subunit of β-conglycinin was identified as an allergenic protein (Ogawa et al. 1995). The α' and β subunits were also identified as potential allergens by immunoblot analysis (Krishnan et al. 2009). Despite the negative factors associated with β-conglycinin, it has several health benefits. It was shown that β-conglycinin is effective in decreasing serum triglyceride levels (Moriyama et al. 2004) and maintaining body fat ratios (Baba et al. 2004). ACE inhibitory activity for rats was identified in peptides from β-conglycinin (Matsui et al. 2003). Protease hydrolyses of the β-conglycinin yielded antioxidative activity against the peroxidation of linoleic acid (Chen et al. 1995) The peptide fragment VRIRLLQRFNKRS of the β subunit of β-conglycinin induced satiety (Nishi et al. 2003). The peptide derived from β-conglycinin, soymetide-4 and soymorphin-5, have an anti-alopecia effect (Tsuruki et al. 2005) and auxiolytic activities (Ohinata et al. 2007) respectively. Therefore, modifying globulin composition and subunit composition of each globulin in soybean breeding is important to both health management and food processing.

1. VARIANTS OF LOW LEVELS OF Β-CONGLYCININ

Kitamura and Kaizuma (1981) identified two mutants, Keburi and Moshidou Gong 503, with reduced β-conglycinin levels. Keburi lacked the α' subunit, while Moshidou Gong 503 showed lower levels of the α and β subunits. Keburi lacks *CG-1* (Ladin et al. 1984, Harada et al. 1989, Yoshino et al. 2002) and *CG-2* (Yoshino et al 2002). Moshidou Gong 503 lacks *CG-3* (Yoshino et al. 2002). Seven lines with low levels of β-conglycinin were developed from a cross Keburi × (Oodate No.1×Moshidou Gong 503) (Ogawa et al. 1989). These lines were characterized by the lack of the α' subunit and a marked decrease of both the α, and β subunits. The marked modification in the protein composition did not adversely affect the total protein contents suggesting that glycinin may compensate for the decrease in β-conglycinin levels (Ogawa et al. 1989). Actually, a high negative correlation was observed between the glycinin and β-conglycinin contents (Ogawa et al. 1989). A line from a cross of Keburi and Moshidou Gong 503 was crossed with Tohoku 80 to produce Karikei 434, a line with α' subunit deletion and lower levels of the α and β subunits. A mutant line completely lacking both the α and α' subunits was isolated from the progeny of Kari-kei 434 seeds treated by γ-ray irradiation (Takahashi et al. 1994). A four base pair insertion in the first exon of wild *CG-2*, which was derived from Moshidou Gong 503, introduced a premature stop codon leading to the α subunit deficiency (Ishikawa et al. 2006). This line is named Yumeminori and released from National Agricultural Research Center for Tohoku Region in 2001 (Takahashi et al. 2004). This cultivar is rich in glycinin and lacks two major allergens, the α subunit and Gly m Bd 28K. The deficiency of Gly m Bd 28K was derived from Tohoku 80. A soybean cultivar, Nagomimaru, was developed from the backcross between Yumeminori and Tachinagaha (recurrent parent) (Hajika et al. 2009). This cultivar has a similar seed protein composition and shows higher yield compared with Yumeminori.

2. CHARACTERIZATION OF Β-CONGLYCININ DEFICIENT MUTANTS

The genes encoding globulin subunits are expressed only in seeds and only during the mid-maturation stage, making them attractive models for the study of mechanisms underlying seed-specific and stage-specific gene expression. Air-dried seeds of the soybean cultivar, Yumeyutaka, were γ-irradiated and a mutant line lacking the α, α' and β subunits of β-conglycinin was isolated (Kitagawa et al. 1991). This deficiency is controlled by a single recessive gene *cgdef* (*β-conglycinin-deficient*) and the mutant has a similar seed protein composition except for β-conglycinin to that of Yumeyutaka. This mutation is associated with *cgdef* syndrome, morphological abnormality, sterility and lethality. So *cgdef* gene can be transferred only by heterozygous state. The genotype of *Cgdef* locus of each seed in the progeny derived from self-crossed wild type plants was identified by sodium dodecyl sulfate (SDS) polyacrylamide gel electrophoresis (SDS-PAGE) (Figure 1). The results of Southern and northern blot analyses indicated that the deficiency is not caused by a lack of, or structural defects in, β-conglycinin subunit genes, but rather arise at the mRNA level (Hayashi et al. 1998). Transient expression of glucuronidase from the promoters of the α' and β subunit genes was detected in the mutant cotyledons suggesting the normal promoter activities of the subunit genes in the mutant (Hayashi et al. 1998). The results of gel mobility shift assays using the 5'-flanking regions of the α' and β subunit genes failed to detect a deficiency of nuclear factors interacting with these regions (Hayashi. et al. 1998). The activation of naked chimeric genes in the mutant seeds might not necessarily indicate the presence of all transcriptional factors that are necessary for activation of β-conglycinin subunit genes in the chromosome. Though the several loci on the two chromosomes encode β-conglycinin subunits, a single recessive gene controls the null trait of the mutant. This, taken together with above results leads to the assumption that a seed-specific mechanism of expression of β-conglycinin subunit genes might be in chromatin organization, and that such an organization might not work normally in the mutant. The possibility that transcript

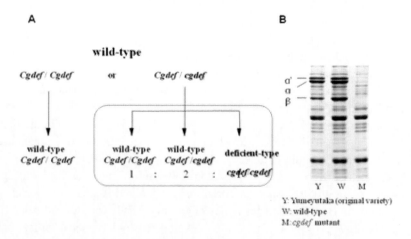

Figure 1. Procedure to get the *cgdef* mutants by self-hybridization of wild type plants and SDS-PAGE. A The *cgdef* mutants and heterozygotes segregate in progeny from self-hybridization of wild type plants. The *cgdef* gene can be maintained only as heterozygotes. B Identification of the *cgdef* mutants by SDS-PAGE using a distal part of each seed.

stability is lowered in the mutant is not excluded. The mutant gene might encode a common factor that regulates the β-conglycinin subunit genes. The two cases are schematically shown in Figure 2. Further genetic analyses revealed that *cgdf* syndrome is derived not from pleiotropy of the *cgdf* gene but from close linkage with some other unidentified genes

(Hayashi et al. 2000). The *Cgdef* locus was mapped on chromosome 19 between simple sequence repeat (SSR) markers Satt523 and Sat_388 (Hayashi et al. 2009).

The study of the *Cgdef* gene might lead to the elucidation of the regulation mechanism of β-conglycinin subunit gene expression in soybean developing seeds and provide us with a clue to modify the soy protein.

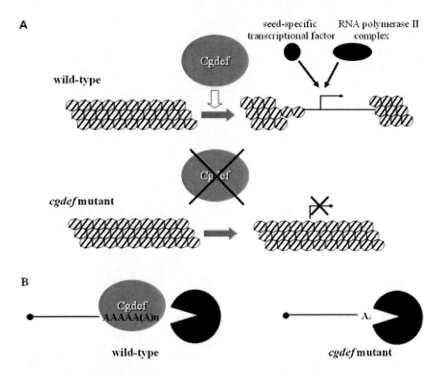

Figure 2. Schematic representation of function of *Cgdef* gene product. A The *Cgdef* gene product acts as a factor that regulate expression of β-conglycinin subunit genes in a chromachin organization manner. B The *Cgdef* gene product functions to stabilize mRNAs ofβ-conglycinin subunits.

Another β-conglycinin deficient mutant QT2 was identified from wild soybeans (Hajika et al. 1996), and the phenotype was found to be controlled by a single dominant gene *Scg-1* (*suppressor of β-conglycinin -1*) (Hajika et al. 1998). *Scg-1* was introduced into a soybean cultivar Fukuyutaka from QT2 and this near-isogenic (NIL) was designated as QY7-25. SDS-PAGE analysis of seed proteins, and detection of β-conglycinin subunit mRNAs in immature seeds by reverse transcription PCR (RT-PCR) with specific primer pairs for α, α' and β subunits were performed using Fukuyutaka and QY7-25 (Figure 3). QY7-25 line lacked all three subunit of β-conglycinin and a much lower amount of β-conglycinin transcripts was amplified from QY7-25 than from Fukuyutaka (Tsubokura et al. 2006a). Two cDNA libraries were constructed from immature seeds of QY7-25 and Fukuyutaka. Screening of the QY7-25 cDNA library revealed the existence of a positive clone, which encode β-conglycinin subunits at a frequency of about one in 10,000 plaques (0.01%). In contrast, the Fukuyutaka cDNA library contained about 5% of β-conglycinin subunit cDNA clones (Tsubokura et al. 2006a). Single nucleotide polymorphisms (SNPs) in the β subunit genes were detected between QY7-25 and Fukuyutaka. Two β subunit genes were found to cosegregate with β-conglycinin deficiency using the DNA marker based on SNPs in an F_2 population from a cross between QY7-25 and Fukuyutaka.. The region associated with β-conglycinin deficiency was located on chromosome 11 with the developed marker using the F_2 population from the parents, Misuzudaizu and Moshidou Gong 503 (Tsubokura et al. 2006b). Later, it was found that the β subunit gene *CG-β-2* (Glyma20g28640) is located on the *Scg-1* region but another β subunit gene, *CG-β-1* (Glyma20g28460.1) recombines with the *Scg-1* region (Tsubokura et al. 2012). The α subunit gene, *CG-α-1* had an SNP in the fourth intron between QY7-25 and Fukuyutaka. Results of the segregation analysis with CAPS marker based on this SNP in the F_2 population derived from QY7-25 and Fukuyutaka showed complete cosegregation with *scg-1* (normal) allele. Therefore at least one α subunit gene and one β subunit gene are located on *Scg-1* region. The physical map of the *Scg-1* region covered by lambda phage genomic clones revealed the two α subunit gene, a β subunit gene, and a pseudo α subunit gene were closely organized (Tsubokura et al. 2012) (Figure 4).

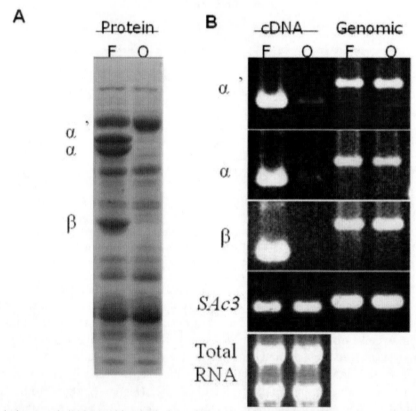

From Tsubokura et al. (2006a) with permission of Springer +Business Media.

Figure 3. SDS- PAGE analysis of β-conglycinin and detection of β-conglycinin mRNA by RT-PCR assay. A Lanes F and Q show the protein extracted from Fukuyutaka and QY7-25, respectively. QY-7-25 lacks all three subunits of β-conglycinin. B The specific primer pairs for α', α and β subunit genes were used. Lanes F and Q show PCR products of Fukuyutaka and QY-7-25, respectively. Since the PCR products amplified from genomic DNA include introns, the bands migrated slower than cDNA. Soybean actin 3 gene (*Sac3*) was used as a control. Total RNA used for reverse transcription is shown at the bottom of the figure.

The two α subunit genes, *CG-α-1* and *CG-α-2*, were arranged in a tail-to-tail orientation, and the genes were separated by 197 bp in *Scg-1* compared to 3.3kb in the wild allele (*scg-1*). The coding sequences of the two α subunit genes were identical in QY7-25. These α subunit genes have tandem polyadenylation signals in normal soybean but a posterior polyadenylation signal was deleted in *CG-α-1* of QY7-25. So read-through transcript may form a stem-loop structure and the structure could be considered the trigger of homology-dependent gene silencing (Figure 5).

RNA silencing consists of three processes, formation of double-strand RNA (dsRNA), processing of dsRNA into small (approximately 21-25 nucleotides) RNAs (sRNA), and a selected sRNA strand within effector complexes guides them to partially complementary target sites of RNA or DNA for inhibitory action (Brodersen and Voinnet 2006). Northern blot analysis was performed to detect the small RNA hallmark of the RNA degradation. As a result, sRNAs were detected in QT2 and QY7-25 with the 3'-end probe only and not with the 5'-end prove, suggesting that the read-through transcript covered only the 3' region of the flip-side gene (Tsubokura et al. 2012). Among the β-conglycinin subunit genes, the 3' regions share particularly high sequence homology. The sRNA could be the guide to RNA silencing for all three subunits. These results strongly suggest that β-conglycinin deficiency in QT2 is controlled by post-transcriptional gene silencing through the inverted repeat of the α subunits. To confirm the read-through transcription, northern blot analysis and RT-PCR were performed, but transcript from the tail- to-tail inverted repeat was not detected (Tsubokura et al. 2012). The read-through transcription may have occurred but with very low frequency, and/or the transcript may have been degraded immediately.

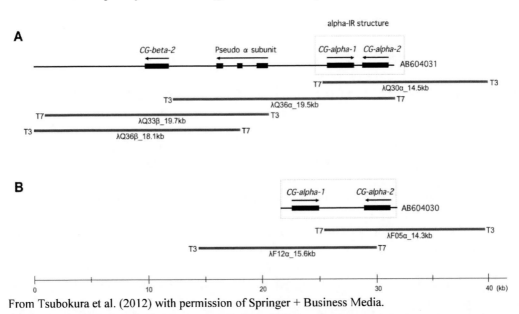

From Tsubokura et al. (2012) with permission of Springer + Business Media.

Figure 4. Physical map for the *Scg-1/scg-1* region of QY7-25 (A) and Fukuyutaka (B). The QY7-25 contig consists of four genomic clones, and the Fukuyutaka contig consists of two genomic clones. The inverted repeat structure is designated as the alpha-IR.

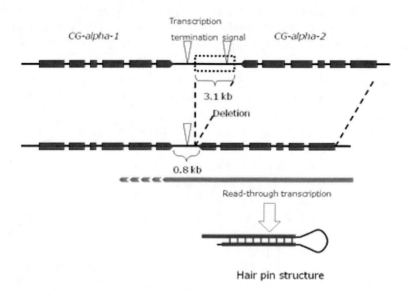

Figure 5. Double-strand RNA with a stem-loop structure from a read-through transcript was predicted at the alpha-IR and schematically displayed.

No agronomic disadvantage in QY7-25 was observed and *Scg-1* gene is a valuable gene for increasing globulin in soybean breeding. The *Scg-1* gene is being introduced into elite cultivars.

4. MUTANTS WITH LOW LEVELS OF GLYCININ CONTENT

The absence of $A_4A_5B_3$ subunit is controlled by a single recessive allele and about 25% of Japanese local varieties lack $A_4A_5B_3$ subunit (Harada et al. 1983, Kitamura et al. 1984). A mutant line lacking all group I ($A_{1a}B_{1b}$, $A_{1b}B_2$, A_2B_{1a}) subunits was induced by γ-irradiated seeds of Wasesuzunari (Odanaka and Kaizuma 1989). Kitamura et al. (1993) found one accession of *Glycine soja* that lacks the A_3B_4 subunit. F_2 seeds from a cross between a mutant line lacking group I and $A_5A_4B_3$ subunits, and a wild soybean (*Glycine soja*) accession lacking A_3B_4 subunit were examined for glycinin subunits by SDS-PAGE. Eight phenotypes including a wild type with all the glycinin subunits, three phenotypes having single deficiencies, three phenotypes having double deficiencies and a triple deficiency lacking all the subunits, appeared among the F_2 seeds tested (Figure 6). The segregation for presence or absence of the glycinine subunits fitted well to a single locus recessive inheritance of each deficiency and independent heredity among the three types of deficiencies (Yagasaki et al. 1996). An F_2 line lacking group I and A_3B_4 subunits was backcrossed with Suzuyutaka lacking $A_5A_4B_3$ subunit to get a line (B2F2-010) lacking group I and A_3B_4 subunits with Suzuyutaka genetic background. Then B2F2-010 was backcrossed with the elite cultivars Tamahomare and Enrei, each lacking $A_5A_4B_3$ subunit. Eight phenotypes described above were classified based on SDS-PAGE at the BCnF2 generation and NILs of the eight phenotypes with genetic backgrounds of Tamahomare and Enrei were obtained (Yagasaki 2011). The NILs of Enrei showed lower seed size but similar yield and protein content compared with Enrei. No significant difference was observed on seed size, yield and protein content between Tamahomare and its NILs (Yagasaki 2011). The content of β-conglycinin of a NIL lacking group I, $A_5A_4B_3$ and A_3B_4 subunits with Enrei background is 1.5 times higher compared with Enrei and 1.9 times higher compared with a line having all the glycinin subunits, suggesting compensation by increase of β-conglycinin for decrease of glycinin (Yagasaki 2011).

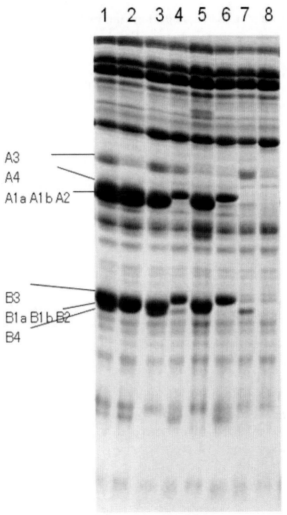

From Yagasaki (2011) with permission of the author and publisher.

Figure 6. SDS-PAG analysis of the total globulin fraction from eight types of induced mutant lines. 1, a wild type with all the glycinin subunits; 2 to 8, mutant lines lacking A_3B_4, $A_4A_5B_3$, group I, A_3B_4 and $A_4A_5B_3$, group I and A_3B_4, group I and $A_4A_5B_3$, all the subunits, respectively.

Using these NILs, glycinin molecule composed of only group I subunit was found to have sedimentation coefficient of 7S and the highest solubility at 0.05M NaCL and pH5.5 compared with other types of glycinin (Yagasaki 2011). The ratios of glycinin/β-conglycinin of soy milks were varied significantly among NILs and the hardness of tofu curds measured by the breaking stress showed high positive correlation with glycinin/β-conglycinin. About the NILs of Enrei, the ratios of glycinin/β-conglycinin of soy milks, and breaking stresses of tofu curds were shown in Table 1 (Yagasaki 2011). It was supposed that the hardness of tofu gel and speed of gelation could be significantly changed by soybean lines with variation of glycinin subunit compositions. When the glycinin content was adjusted to an identical level in soy milks having group I, A_3B_4 and $A_5A_4B_3$ subunits, the order of tofu card hardness based on breaking stress was $A_5A_4B_3$, A_3B_4 and group I (Tezuka et al. 2000).

Table 1. The ratio of glycinin/β-conglycinin and breaking stress of tofu-gel in the NILs of Enrei

Line	Glycinin Composition[1]			The ratio of 11S/7S[2] in soy milk	Breaking stress of tofu-gel (Pa)
	I	IIa	IIb		
EnB2-111	+	+	+[3]	1.97	9989
EnB2-110	+	+	–	1.61	8955
EnB2-101	+	–	+	1.33	10171
EnB2-011	–	+	+	0.82	7162
EnB2-100	+	–	–	1.09	6791
EnB2-010	–	+	–	0.48	4835
EnB2-001	–	–	+	0.33	5381
EnB2-000	–	–	–	0.13	3002
Enrei	+	–	+	1.36	9891

[1] I, $A_{1a}B_{1b}$, $A_{1b}B_2$; IIa, $A_4A_5B_3$; IIb, A_3B_4
[2] 11S, glycinin ; 7S, β-conglycinin
[3] + , presence ; – , absence
The table was selected from (Yagasaki 2011) and modified with permission
of the author and publisher.

A soybean cultivar named Nanahomare lacking group I, $A_5A_4B_3$ and A_3B_4 subunits was developed in the progeny from a cross between a NIL (TmB4-010) of Tamahomare lacking group I and A_3B_4 subunit and Tamahomare lacking $A_5A_4B_3$ (Yagasaki et al. 2010). Nanahomare displayed similar growth characteristics, yield and seed quality including protein content and 1.8-fold increase of β-conglycinin in its seeds compared with Tamahomare. Hardness of tofu curds of Nanahomare was greatly reduced compared with Fukuyutaka or Tachinagaha due to glycinin deficiency. Because β-conglycinin is effective in decreasing serum triglyceride levels and maintaining body fat ratios, new food processing for Nanahomare is necessary.

Nanahomare is an effective source for isolating β-conglycinin. As the protein content of Nanahomare is relatively low (about 42%), a breeding program to increase protein content is under way at Nagano Prefecture Vegetable and Ornamental Crops Experimental Station.

5. A LINE THAT LACKS BOTH GLICININ AND B-CONGLYCININ

A line QF2 was developed from a cross between QY2 and EnB1. QT2 is a β-conglycinin deficient mutant described above. EnB1 is a NIL of Enrei that lacks all the glycinin subunits. QF2 lacks all major subunits of glycinin and β-conglycinin (Takahashi et al. 2003). In spite of the absence of storage protein subunits, QT2 grew normally and the nitrogen content of its seeds was similar to that of wild-type varieties. But protein bodies underdeveloped in the cotyledons of QF2. Although total free amino acids contributed only 0.3-0.8% of the nitrogen content of seeds of wild-type varieties, they contributed 4.5-8.2% of the nitrogen content of seeds in the QF2 line, with arginine being especially enriched. It was suggested that the QF2 line maintains its seed nitrogen content by accumulating free amino acids in addition to increasing the abundance of certain seed proteins including lipoxygenase sucrose binding protein, lectin, P34 putative thiol protease, basic 7S globulin, and so on (Takahashi et al. 2003). As the seeds of QF2 contain 0.6-1.5% free arginine, it could be used directly or indirectly as a dietary source of this amino acid. The QF2 line also may be amenable to the accumulation foreign proteins by transformation for biotechnological needs.

CONCLUSION

Glycinin and β-conglycinin show different physicochemical functions, including gelation, emulsification and foaming, and different physiological functions, including ACE inhibitory activity, reduction of serum triglyceride levels, antioxidative activity and serum choresterol lowering effect. Furthermore, the subunit composition of glycinin has remarkable effects on characters of gels and gelation process. The mutants of β-conglycinin with α'-less, lower levels of α, β subunits, α and α'-less, α, α' and β-less were already available. The mutants of glycinin with $A_4A_5B_3$-less, group I subunit-less, A_3B_4-less, three phenotypes having double deficiencies and a triple deficiency lacking all the subunits were also reported. Furthermore, the line that lacks both glycinin and β-conglycinin was developed. It has become possible to manipulate genetically globulin composition and subunit compositions of both glycinin and β-conglycinin for special uses of food processing or health benefits.

Further works are necessary for isolating the *Cgdef* gene to reveal the regulation mechanism of β-conglycinin subunit gene expression. Molecular genetic analyses on globulin subunit deficiencies is required especially group I subunit deficiency. Though the group I subunit deficiency is controlled by a single recessive gene, $A_{1a}B_{1b}$ and A_2B_{1a} subunits are encoded by *Gy1* and *Gy2* respectively which are located in tandem on chromosome 3 and A_1B_2 subunit is encoded by *Gy3* on chromosome 19. The group I subunit deficiency may be regulated by a defect of a common trans-acting factor for the *Gy1*, *Gy2* and *Gy3* genes as is discussed by Beilinson et al. (2002). Identification of this factor might lead to solution of the regulation of group I subunit gene expression.

REFERENCES

Baba, T., A. Ueda, M. Kohno, K. Fukui, C. Miyazaki, M. Hirotsuka and M. Ishinaga (2004) Effects of soybean β-conglycinin on body fat ratio and serum lipid levels in healthy volunteers of female university students. *J. Nutr. Sci. Vitaminol.* 50:26-31.

Beilinson, V., Z. Chen, R.C. Shoemaker, R.L. Fischer, R.B. Goldberg and N.C. Nielsen (2002) Genomic organization of glycinin genes in soybean. *Theor. Appl. Genet.* 104:1132-1140.

Bordersen, P. and O. Voinnet (2006) The diversity of RNA silencing pathways in plants. *Trends Genet.* 22:268-280.

Chen, H.-M., K. Muramoto and F. Yamauchi (1995) Structural analysis of antioxidative peptides from soybean β-conglycinin. *J. Agric. Food Chem.* 43:574-578.

Fukushima, D. (1968) Internal structure of 7S and 11S globulin molecules in soybean proteins. *Cereal Chem.* 45:203-224.

Fukushima, D. (1991) Structure of plant storage proteins and their functions. *Food Reviews International* 7: 353-381.

Hajika, M., M.Takahashi, S.Sakai and M. Igita (1996) A new genotype of 7S globulin (β-conglycinin) detected in wild soybean (*Glycine soja* Sieb. et Zucc.). *Breed. Sci.* 46:385-386.

Hajika, M., M. Takahashi, S. Sakai and R. Matsunaga (1998) Dominant inheritance of a trait lacking β-conglycinin detected in a wild soybean line. *Breed. Sci.* 48:383-386.

Hajika, M., K. Takahashi, T. Yamada, K. Komaki, Y. Takada, H. Shimada, T. Sakai, S. Shimada, T. Adachi, K. Tabuchi, A. Kikuchi, S. Yumoto, S. Nakamura and M. Ito (2009) Development of a new soybean cultivar for soymilk, "Nagomimaru". *Bull. Natl. Inst. Crop Sci.* 10:1-20.

Harada, J.J, S.J. Barker and R.B. Goldberg (1989) Soybean β-conglycinin genes are clustered in several DNA regions and are regulated by transcriptional and posttranscriptional processes. *Plant Cell* 1:415-425.

Harada K., Y. Toyokawa and K. Kitamura (1983) Genetic analysis of the most acidic 11S globulin subunit and related characters in soybean seeds. *Japan. J. Breed.* 33:23-30.

Hayashi, M., K. Harada, T. Fujiwara and K. Kitamura (1998) Characterization of a 7S globulin-deficient mutant of soybean (*Glycine max* (L.) Merrill). *Mol. Gen. Genet.* 258:208-214.

Hayashi, M., M. Nishioka, K. Kitamura and K. Harada (2000) Identification of AFLP markers tightly linked to the gene for deficiency of the 7S globulin in soybean seed and characterization of abnormal phenotypes involved in the mutation. *Breed. Sci.* 50: 123-129.

Hayashi, M. K. Kitamura and K. Harada (2009) Genetic mapping of *Cgdef* gene controlling accumulation of 7S globulin (β-conglycinin) subunits in soybean seeds. *J. Hered.* 100:802-806.

Ishikawa,_G., Y. Takada and T. Nakamura (2006) A PCR-based method to test for the presence or absence of β-conglycinin α'- and α-subunits in soybean. *Mol. Breed.* 17:365-374.

Kitagawa, S., M. Ishimoto, F. Kikuchi and K. Kitamura (1991) A characteristic lacking or decreasing remarkably 7S globulin subunits induced with γ-ray irradiation in soybeam seeds. *Jpn. J. Breed.* 41 (Suppl. 2):460-461.

Kitamura, K., and N. Kaizuma (1981) Mutant strains with low level of 7S globulin in soybean (*Glycine max* Merr.) seed. *Japan. J. Breed.* 31:353-359.

Kitamura, K., C.S. Davis and N.C. Nielsen (1984) Inheritance of alleles of *Cgy1* and *Cgy4* storage protein genes in soybean. *Theor. Appl. Genet.* 68:253-257.

Kitamura, K., M. Ishimoto and N. Kaizuma (1993) Genetic relationships among genes for the subunits of soybean 11S globulin. *Jpn. J. Breed.* 43 (Suppl. 2) 159.

Kohyama, K. and K. Nishinari (1993) Rheological studies on the gelation process of soybean 7S and 11S proteins in the presence of Glucono-δ-lactone. *J. Agric. Food Chem.* 41:8-14.

Krishnan, H.B., W.S, Kim, S. Jang and M.S. Kerley (2009) All three subunits of soybean β-conglycinin are potential food allergens. *J. Agric. Food Chem.* 57:938-943.

Kuba, M., K. Tanaka, S. Tawata, Y. Takeda and M. Yasuda (2003) Angiotensin I-converting enzyme inhibitory peptides isolated from tofuyo fermented soybean food. *Biosci. Biotechnol. Biochem.* 67:1278-1283.

Ladin, B.F., J.J. Doyle and R.N. Beachy (1984) Molecular characterization of a deletion mutation affecting the α'-subunit of β-conglycinin of soybean. (1984) *J. Mol. Appl. Genet.* 2:372-380.

Li, C. and Y.-M. Zhang (2011) Molecular evolution of *glycinin* and *β-conglycinin* gene families in soybean (*Glycine max* L. Merr.). *Heredity* 106:633-641.

Mallikarjun Gouda, K.G., L.R. Gowda, A.G. Appu Rao and V. Prakash (2006) Angiotensin I-converting enzyme inhibitory peptide derived from glycinin, the 11S globulin of soybean (*Glycine max*). *J. Agric. Food Chem.* 54:4568-4573.

Matsui, T. (2003) Production of hypotensive peptide, SVY, from 7S globulin of soybean protein and its physiological functions. *Soy Protein Research,* Japan 6:73-77.

Moriyama, T., K. Kishimoto, K. Nagai, R. Urade, T. Ogawa, S. Utsumi, N. Maruyama and M. Maebuchi (2004) Soybean β-conglycinin diet suppresses serum triglyceride levels in normal and genetically obese mice by inducing of β-oxidation, downregulation of fatty acid synthase, and inhibition of triglyceride absorption. *Biosci. Biotechnol. Biochem.* 68:352-359.

Murasawa, H., A. Sakamoto, H. Sasaki and K. Harada (1991) The effect of a glycinin subunit on tofu-making. Proceeding of the International Conference On Soybean Processing And Utilization pp. 53-57.

Nakamura, T., S. Utsumi, K. Harada and T. Mori (1984) Cultivar differences in gelling characteristics of soybean. *J. Agric. Food Chem.* 32:647-651.

Nakamura, T., S. Utsumi and T. Mori (1985) Formation of pseudo-glycinins from intermediary subunits of glycinin and their gel properties and network structure. *Agric. Biol. Chem.* 49:2733-2740.

Nielsen, N.C., R. Bassuner and T. Beaman (1997) The biochemistry and cell biology of embryo storage proteins. In: Larkins, B.A. and I. K. Vasil (ed) Cellular and Molecular Biology of Plant Seed Develpoment pp.151-220, Kluwer Academic Publishers, Dordrecht.

Nishi, T., H. Hara, K. Asano and F. Tomita (2003) The soybean β-conglycinin β 51-63 fragment suppresses appetite by stimulating cholecystokinin release in rats. *J. Nutr.* 133: 2537-2542.

Nishinari K., K. Kohyama, Y. Zhang, K. Kiyamura, T. Sugimoto, K. Saio and Y. Kawamura (1991) Rheological study on the effect of the A_5 subunit on the gelation characteristics of soybean proteins. *Agric. Biol. Chem.* 55:351-355

Odanaka H. and N. Kizuma (1989) Mutants on soybean storage proteins induced with γ-ray irradiation. *Jpn. J. Breed._*39 (Suppl. 1) 430-431.

Ogawa, T., N. Bando, H. Tsuji, K. Nishikawa and K. Kitamura (1995) α-Subunit of β-conglycinin, an allergenic protein recognized by IgE antibodies of soybean-sensitive patients with atopic dermatitis. *Biosci. Biotechnol. Biochem._*59:831-833.

Ogawa, T., E. Tayama, K. Kitamura and N. Kaizuma (1989) Genetic improvement of seed storage proteins using three variant alleles of 7S globulin subunits in soybean (*Glycine max* L.). *Japan. J. Breed.* 39:137-147.

Ohinata, K., S. Agui and M. Yoshikawa (2007) Soymorphins, novel μ opioid peptides derived from soy β-conglycinin β-subunit, have anxiolytic activities. *Biosci. Biotechnol. Biochem.* 7:2618-2621.

Saio, K., M. Kamiya and T. Watanabe (1969) Food processing characteristics of soybean 11S and 7S proteins. Part I. Effect of difference of protein components among soybean varieties on formation of tofu-gel. *Agr. Biol. Chem.* 33:1301-1308.

Saio, K., M. Kamiya and T. Watanabe (1971) Food processing characteristics of soybean 11S and 7S proteins. Part II. Effect of sulfhydryl group on physical properties of tofu-gel. *Agric. Biol. Chem.* 35:890-898.

Takahashi, K., H. Banba, A. Kikuchi, M. Ito and S. Nakamura (1994) An induced mutant line lacking the α-subunit of β-conglycinin in soybean [*Glycine max* (L.) Merrill]. *Breed. Sci.* 44:65-66.

Takahashi, M., Y. Uematsu, K._Kashiwaba, K. Yagasaki, M. Hajika, R. Matsunaga, K. Komatsu and M. Ishimoto (2003) Accumulation of high levels of free amino acids in soybean seeds through integration of mutations conferring seed protein deficiency. *Planta* 217:577-586.

Takahashi, K., S. Shimada, H. Shimada, Y. Takada, T. Sakai, Y. Kono, T. Adachi, K. Tabuchi, A. Kikuchi, S. Yumoto, S. Nakamura, M. Ito, H. Banba and A. Okabe. (2004) A new soybean cultivar "Yumeminori" with low allergenicity and high content of 11S globulin. Bull. Natl. Agric. Res. Cent. Tohoku Reg. 102:23-39.

Takahashi, K. (2010) Breeding of a new soybean cultivar "Nanahomare". *The Hokuriku Crop Science* 45:61-64.

Tezuka, M., H. Taira, Y. Igarashi, K. Yagasaki and T. Ono (2000) Properties of tofus and soy milks prepared from soybeans having different subunits of glycinin. *J. Agric. Food Chem.* 48:1111-1117.

Tsubokura,_Y., M. Hajika and K. Harada (2006a) Molecular characterization of a β-conglycinin deficient soybean. *Euphytica* 150:249-255.

Tsubokura, Y., M. Hajika and K. Harada (2006b) Molecular markers associated with β-conglycinin deficiency in soybean. *Breed. Sci.* 56:113-117.

Tsubokura, Y., M. Hajika, H. Kanamori, Z. Xia, S. Watanabe, A. Kaga, Y. Katayose, M. Ishimoto and K. Harada (2012) The β-conglycinin deficiency in wild soybean is associated with the tail-to-tail inverted repeat of the α- subunit genes. *Plant Mol. Biol.* 78:301-309.

Tsuruki, T., K. Takahata and M. Yoshikawa (2005) Anti-alopecia mechanisms of soymetide-4, an immunostimulating peptide derived from soy β-conglycinin. *Peptides* 26:707-711.

Urade, R. and M. Kohno (2011) Physiological function of highly purified soy proteins. *Soy Protein Research*, Japan 14: 7-11.

Utsumi, S. and J.E. Kinsella (1985) Forces involved in soy protein gelation :Effects of various reagents on the formation, hardness and solubility of heat-induced gels made from 7S, 11S, and soy isolate. *J. Food Sci.* 50:1278-1282.

Yagasaki, K., N. Kaizuma and K. Kitamura (1996) Inheritance of glycinin subunits and characterization of glycinin molecules lacking the subunits in soybean (*Glycine max* (L.) Merr.). *Breed. Sci.* 46: 11-15.

Yagasaki, K., F. Kousaka and K. Kitamura (2000) Potential improvement of soymilk gelation properties by using soybeans with modified protein subunit compositions. *Breed. Sci.* 50:101-107.

Yagasaki, K., H. Sakamoto, K. Seki, N. Yamada, M. Takamatsu, T. Taniguchi and K. Takahashi (2010) Breeding of a new soybean cultivar "Nanahomare". *The Hokuriku Crop Science* 45:61-64.

Yagasaki, K. (2011) Studies on genetic improvement of glycinin subunit composition in soybean seed storage proteins. *Special Bull. Nagano Veg. & Ornam. Crops Exp. Sta. Japan* 3:1-56.

Yoshino, M., A. Kanazawa, K. Tsutsumi, I. Nakamura, K. Takahashi and Y. Shimamoto (2002) Structural variation around the gene encoding the α subunit of soybean β-conglycinin and correlation with the expression of the α subunit. *Breed. Sci.* 52:285-292.

In: Agricultural Research Updates. Volume 5
Editors: P. Gorawala and S. Mandhatri

ISBN: 978-1-62618-723-8
© 2013 Nova Science Publishers, Inc.

Chapter 6

DIFFERENT EFFECTS OF GARLIC PREPARATIONS: A MINI REVIEW

Luziane P. Bellé[1] and Maria Beatriz Moretto[1]*

[1]Postgraduate Program in Pharmacology
Health Sciences Center
Federal University of Santa Maria, Santa Maria, RS, Brazil

ABSTRACT

Currently, people look for alternative methods for the treatment of diseases, especially those coming from the aging. Garlic (Allium sativum), is a plant known since ancient times that has been used to treat a large number of diseases in popular medicine. There are different types of preparations (raw, extracts, oil and powder), and each one has specific effects according to the predominant components. These compounds stem from the way that the preparation is obtained. Therefore, this study was aimed at discussing the beneficial and side effects of the use of garlic preparations and garlic bulbs in the alimentation.

Keywords: Garlic; garlic preparations; popular medicine

INTRODUCTION

Medicinal plants continue to provide valuable therapeutic agents, in both modern medicine and in traditional system (El-Demerdash, 2005). Ethnobotanical research has increased considerably in recent years, and it has been considered a subject of great interest. They have played a central role in the prevention and treatment of diseases since prehistoric times.

It seems therefore paradoxically that, of the many hundreds of plant remedies described in the Papyrus of Ebers, discovered in Egypt and written in about 1550 BC, only a handful of

---

* Corresponding author: Luziane Potrich Bellé, luzi_belle@yahoo.com.br, Fax: +55 - 3220 8018.

products currently appear among the accepted remedies in the United States Pharmacopoeia (Talalay and Talalay 2001). Of particular interest are the plants used in popular medicine, for some of which it has been possible to confirm what tradition has reported. Many remedies suggested by the ancients have also been included, through scientific demonstration, in official medicine (Lentini, 2000).

Garlic (Allium sativum L., Liliaceae) is a common spicy flavoring agent used since ancient times. It has been used therapeutically from the earliest records, nearly 5,000 years ago. There is evidence that Chinese used garlic around 3,000 years back (Tattelman, 2005). The plant has been cultivated in all over Iran because of its characteristic flavor and medicinal properties.

Although garlic has been used for centuries, and even nowadays is part of popular medicine in many cultures, few scientific reports on its therapeutic and pharmacological properties are found. In the past decade, some protective effects of garlic have been well established by epidemiological studies and animal experiments and have validated many of the medicinal properties attributed to garlic and its potential to lower the risk of disease (Eidi, 2006, Borek, 2001).

A. sativum contains various biologically active constituents, as alliin, alliinase, allicin, S-allycystein, diallylsulphide and allymethyltrisulphide. Alliin is an amino acid, which is converted into allicin by alline-lyase catalyses when the bulbs are crushed (Groppo et al., 2007). This transformation is completed in seconds and the alliinase is present in unusually high amounts in garlic cloves: at least 10% of the total protein content (Ankri and Mirelman, 1999).

Allicin is the precursor of sulphur-containing compounds, which are responsible for the flavour, odour and pharmacological properties (Groppo et al., 2007). Besides the organosulfur compounds, garlic also contains carbohydrates (e.g. fructans), enzymes (e.g. alliinase, catalase), proteins and free amino acids (e.g. arginine), lipids, polyphenols and phytosterols. Garlic also contains high levels of saponins, phosphorus, potassium, sulfur, zinc, moderate levels of selenium and vitamins A and C and low levels of calcium, magnesium, sodium, iron, manganese and B-complex vitamins. Nearly all of these compounds present in garlic are water-soluble (97%) with a small amount of oil-soluble compounds (for review see, Rahman, 2003).

Numerous commercially processed garlic forms, which differ in the content of bioactive compounds, especially sulphuric, are available on the market. The knowledge of the types of bioactive substances present in garlic and its products, their changes during treatment and pro-health influence is of crucial importance to the diet supplement producers, doctors, pharmacists and consumers (Rahman, 2003). There are dozens of brands of garlic products on store shelves that provide a convenient way to obtain the health benefits of garlic. They can be classified into four groups: garlic essential oil, garlic oil macerate, garlic powder and garlic extract.

The various forms also differ in their ingredients, effects and toxicities. Garlic products that contain the most safe, effective stable and odorless components are the most valuable as dietary supplements (Amagase, 2006).

With the increasing seek for alternative treatments for curing diseases, this review was aimed at showing the different forms of garlic products available and their main purposes, accordingly to the predominant constituent in the preparation.

GARLIC EXTRACTS

Alcoholic and aqueous garlic extracts contain a complex mixture with relatively high concentrations of water-soluble compounds and S-allyl-L-cysteine (SAC) as the most abundant organosulfur compound (Amagase, 2006; Cruz et al., 2007).

SAC has been studied due its protection against oxidation, free radicals, cancer and cardiovascular diseases (Amagase, 2006). Aged garlic extract (AGE) is an odorless garlic preparation widely studied due to its high antioxidant activity and its health-protective potential. In AGE, garlic is aged for up 20 months and the cold-process aging gently modifies harsh and irritating compounds from the raw garlic and naturally generates unique and beneficial compounds through both enzymatic and natural chemical reactions (Aguilera et al., 2010). On the other hand, the aging process is not present in other alcoholic and aqueous garlic extracts.

Antioxidant Effects of Garlic Extracts

Oxidative modification of DNA, proteins, lipids and small cellular molecules by reactive oxygen species (ROS) plays a role in a wide range of common diseases and age-related degenerative conditions. Antioxidants, including those in AGE, lower the risk of injury to vital molecules and in varying degrees may help prevent the onset and progression of disease. The antioxidative actions of AGE and its components are determined by their ability to scavenge ROS and inhibit the formation of lipid peroxides (Borek, 2001). On the other hand, in vivo effects of garlic preparations may derive not only from direct radical scavenging, but also from induction of the endogenous antioxidant defenses (Amorati and Pedulli, 2008). AGE increases cellular glutathione (GSH) in a variety of cells, including those in normal liver and mammary tissue. The ability of AGE to increase glutathione peroxidase and other ROS scavenging enzymes is important in radioprotection and UV suppression of certain forms of immunity, in reducing the risk of radiation and chemically induced cancer and in preventing the range of ROS-induced DNA, lipid and protein damage implicated in the disease and aging processes (Borek, 2001). Rahman (2003) evaluated the effect of the dietary supplementation with AGE for 14 days on the plasma and urine levels of 8-iso-Prostaglandin F2 alpha, a sensitive and specific indicator of lipid peroxidation, in smoking and nonsmoking men and women. The dietary supplementation reduced plasma and urine concentrations of 8-iso-Prostaglandin F2 alpha by 29 and 37% in nonsmokers and by 35 and 48% in smokers. The authors also observed that the antioxidant capacity of non-smokers was approximately twice that of smokers. Moreover, plasma antioxidant capacity of smokers had significantly increased by 53% following supplementation with AGE for 14 days. The results indicate that dietary supplementation with garlic may improve antioxidant status even in the elderly, as they tend to have higher levels of oxidative stress.

Low density lipoprotein (LDL) oxidation accelerates the growth of fatty streaks in blood vessel walls and the formation of plaque, increasing the risk of atherosclerosis, cardiovascular and cerebrovascular disease. Ide and Lau (2001) demonstrated that AGE and its major compound SAC can protect endothelial cells from oxidized LDL-induced injury, resulting in increased cell viability and inhibited thiobarbituric acid reactive substances (TBARS) formation. Also, the effect of garlic extract on acetaminophen-induced oxidative stress in rat hepatocytes was demonstrated by Anoush et al. (2009), who observed that the garlic extract prevented the ROS formation and GSH depletion. Thus, it is observed that the antioxidant effect is provided by AGE and other garlic extracts, which show capacity to stimulate antioxidant system and act as a scavenging agent.

Anti-Cancer Effects

Several epidemiological studies have demonstrated that the garlic extract shows anticancer activity. The consumption of garlic has been reported to reduce carcinogen-induced mammary, colon, lung, stomach, skin and liver cancers (Youn et al., 2008). AGE exerts its anti-cancer action in different and complementary ways, due to the variety of compounds present in the extract such as water and lipid-soluble organosulfur compounds, phenolic compounds, notably allixin, saponins and selenium. Garlic can also inhibit carcinogenesis by modulating carcinogen metabolism and decreasing DNA carcinogen binding. In addition, it may also prevent ROS-induced DNA damage since garlic has strong antioxidant properties (Rahman, 2003).

A recent study showed that the garlic compounds, such as SAC, diallyl disulfide and allicin, inhibit the nuclear factor kappa B (NF-κB) activation. This effect suggests that the anticancer effects of garlic and its sulfur-containing compounds may be mediated through the modulation of NF-κB activation (Youn et al., 2008) once this transcription factor is involved in cancer progression (Vlasova and Moshkovskii, 2006). Gapter et al. (2008) showed that SAC significantly reduced anchorage-dependent and -independent growth of MDA-MB-231 breast tumor cells in a dose- and time-dependent fashion, and sub-lethal SAC-treatment altered mammary tumor cell adhesion and invasion through components of the extracellular matrix. Moreover, diallyl trisulfide treatment inhibited migration and causes apoptosis induction in breast cancer cells by ROS generation and redox-sensitive changes in EMT markers and induces apoptotic cell death in human primary colorectal cancer (Chandra-Kuntal et al., 2013, Yu et al., 2012). Also, fresh extracts of garlic arrested the growth and altered the morphology of breast cancer cells (Modem et al., 2012). This brief explanation suggests that the garlic extract can provide anti-cancer effects by different mechanisms.

Prevention of Cardiovascular Diseases

The main cause of mortality from dangerous coronary atherosclerosis is the coronary artery thrombosis, which leads to myocardial infarctions that in many cases can be fatal (Gorinstein et al., 2006). Seo et al. (2012) suggested that AGE reduce cardiovascular risk factors. In fact, AGE is able to reduce blood pressure, inhibit platelet aggregation and adhesion (the first step in the pathogenesis of any thrombotic episode) lower LDL and elevate HDL cholesterol preventing cardiovascular diseases (Allison et al., 2006; Borek, 2006).

Allison et al. (2006) showed that in the presence of AGE, inhibition of platelet aggregation occurred in a concentration-dependent manner and achieved significance between 1.56–25% (v:v). The authors suggested that the water-soluble compounds present in

AGE are responsible for its antiplatelet aggregation reaction. In another study, the aggregation was stimulated by using three different platelet agonists: epinephrine, ADP and collagen. AGE inhibited the platelet aggregation, but the inhibitory effect was selective, affecting collagen- and epinephrine-induced aggregation more than that stimulated by ADP. It was interesting to note that the inhibitory effect of AGE on platelet aggregation was not strictly dose dependent (Steiner and Li, 2001). Thromboxane B2 (TXB2) is a vasoconstrictor and platelet aggregating agent and thus is potentially thrombotic. Thomson et al. (2000), using a model where different concentrations of aqueous garlic were administered as single doses in the ear vein of rabbits and the TXB2 level was measured in the serum, observed that garlic inhibited the thrombin-induced platelet synthesis of TXB2 in a dose-and time-dependent manner. Moreover, AGE is able to improve vascular elasticity and endothelial function (Larijani et al., 2013). Hypertension is a known risk factor for cardiovascular morbidity and mortality, affecting an estimated one billion individuals worldwide. The garlic extract has also been shown to modulate the production and function of both endothelium-derived relaxing and constricting factors in rat isolated pulmonary arteries, and to reduce blood pressure in hypercholesterolemic subjects (Rahman, 2003). Garlic has also been used to prevent the development of arteriosclerosis, reduce high blood pressure and to improve prothrombin, thrombin and partial thromboplastin times (Ohaeri and Adoga, 2006). A study using SAC (200 mg/kg i.p.) and garlic extract (1.2 ml/kg i.p.) at 30 day intervals showed that both preparations reduced the hypertension and renal damage in nephrectomized rats. This suggest that the antihypertensive and renoprotective effects of SAC and garlic extract are associated with their antioxidant properties, and that they may be used to ameliorate hypertension and delay the progression of renal damage (Cruz et al., 2007). In patients with uncontrolled hypertension, AGE was able to reduce de blood pressure and authors saw that it may be considered as a safe adjunct treatment to conventional antihypertensive therapy (Ried et al., 2013). The cardiovascular-protective effects of garlic have also been evaluated extensively in recent years. S-methylcysteine sulfoxide, a component of garlic, has been shown to reduce blood cholesterol and the severity of atherosclerosis (Ali et al., 2000). Yeh and Liu (2001) treated cultured rat hepatocytes with [2-14C] acetate in the presence or absence of water-soluble garlic compounds at 0.05–4.0 mmol/L for measurement of cholesterol synthesis. Among water-soluble compounds, S-alk (en)yl cysteines (i.e., SAC, S-ethyl cysteine and S-propyl cysteine) exhibited dose-dependent inhibition on the rate of cholesterol synthesis, with maximum inhibition of 40–60% achieved at the concentration of 2.0–4.0 mmol/L. In another study, to identify the principal site of inhibition in the cholesterolgenic pathway and the active components of garlic, cultured hepatoma cells were treated with aqueous garlic extract or its chemical derivatives, and radiolabeled cholesterol and intermediates were identified and quantified. Garlic extract reduced cholesterol synthesis by up to 75% without evidence of cellular toxicity and increased the plasma levels of HDL (Singh and Porter, 2006, Lee et al., 2012).

GARLIC OIL AND THE ANTIBACTERIAL ACTIVITY

Garlic essential oil (GO), usually obtained by distillation of the raw material, is dominated by sulfides (diallyl, allyl methyl and dimethyl mono- to hexasulfides). These products are produced as a result of the degradation of allicin under the harsh thermal treatment in the hydrodistillation procedure (Block E, 1992). Several researchers reported that GO and its diallyl constituents possess antibacterial activity (Ross et al., 2001; Tsao and Yin, 2001; Chung et al., 2007, Velliyagounder et al., 2012) and this effect is due to allicin (Ankri and Mirelman, 1999; Ross et al., 2001).

Ross et al. (2001) found that for a range of gram-positive and gram-negative enteric bacteria, including food-borne pathogens, minimal inhibitory concentrations obtained for GO indicated growth inhibition of all the bacteria tested. In an in vitro study using diallyl monosulphide, dially disulphide, diallyl trisulphide, diallyl tetrasulphide standards and two essential oils against Staphylococcus aureus, methicillin-resistant S. aureus (MRSA) and fungi, (Tsao and Yin, 2001) found that both diallyl monosulphide and dially disulphide showed activity against the MRSA and the fungi tested. Ankri and Mirelman (1999) verified the ability of allicin to react with a model thiol compound (L-cysteine) to form the S-thiolation product, S-allylmercaptocysteine, thus concluding that the main antimicrobial effect of allicin is due to its interactions with important thiol-containing enzymes. In another study, time course viability experiments assessed the anti-H. pylori activity of GO (16 and 32 μg/ml-1) in simulated gastric environments and they observed rapid anti-H. pylori action of GO in artificial gastric juice (O'Gara et al., 2008). Also, diallyl disulphide acts as potent antifungal agent (Yousuf et al., 2010).

On the other hand, the antibacterial properties of crushed garlic have been known for a long time. Various garlic preparations (not only GO and their compounds) have been shown to exhibit a wide spectrum of antibacterial activity against Gram-negative and Gram-positive bacteria (Ankri and Mirelman, 1999). In an in vitro study, the number of viable Acanthamoeba castellanii trophozoites decreased with exposure to increasing concentrations of garlic extract, and generally with increased times of exposure (Polat et al., 2008). Groppo et al. (2007), using saliva samples from thirty healthy subjects for obtaining mutans streptococci, demonstrated that the streptococci strains were susceptible to the garlic extracts. The efficacy in reducing the streptococci in the in vivo assay provided evidence to the susceptibility of oral streptococci, with anticariogenic activity at the concentration of 2.5%.

Garlic Powder

Garlic powder is thought to retain the same ingredients as raw garlic, although the proportions and amounts of various constituents may differ significantly (Sobenin et al., 2008). Dehydrated-garlic powder products contain oil-soluble compounds derived from allicin, and previous reports indicated that these compounds stimulate cytochrome P450 isoenzymes (Amagase, 2006) therefore influencing in hepatic detoxification and xenobiotic metabolism.

Sheen et al. (1996), studying the effect of garlic diallyl sulfide on the detoxification capability and the antioxidation system of primary rat hepatocytes, found that when hepatocytes were treated with diallyl sulfide at 0.5 or 1mM, the intracellular GSH levels were 8-23% higher than in the controls at 24 h, and a significant difference was observed for 1mM treatment at 48 h. This phenomenon is beneficial to the detoxification and antioxidation capabilities of hepatocytes, because GSH protects cells from the toxic effects of reactive oxygen compounds such as free radicals and peroxides.

Potassium dichromate (K2Cr2O7) is a chemical compound that has been demonstrated not only to induce oxidative stress, but it also has carcinogenic potential in nature. Pedraza-Chaverri et al. (2007) induced tubule interstitial damage in rats with a single injection of $K_2Cr_2O_7$ (15 mg/kg). However, garlic powder diet was able to prevent by 44-71% the alterations in renal injury markers.

Cis-diamminedichloroplatinum (II), or cisplatin, is an effective chemotherapeutic agent successfully used in the treatment of a wide range of tumors.

Nevertheless, nephrotoxicity has restricted its clinical use. In an experiment using rats fed a 2% garlic powder diet for 4 weeks, a single injection of cisplatin (7.5 mg/kg) induced tubular damage and increased the levels of blood urea nitrogen, serum creatinine and urinary excretion of N-acetyl-beta-D-glucosaminidase three days post-injection (Razo-Rodríguez et al., 2008).

Therefore, garlic powder was able to prevent by 40-59% the alterations in the renal injury markers studied.

In another study, rats were fed a 5% garlic powder diet and activities of several hepatic enzymes, which are important in carcinogen metabolism such as cytochrome P450 (CYP) and phase II enzymes, were determined. An increase in some enzymatic activities, depending on the particular form of CYP, was observed. Garlic consumption decreased the total CYP concentration in the liver, the level of CYP 2E1 isoform and the PNPH (a marker of CYP2E1 which is able to metabolize nitrosamines and numerous low molecular weight chemicals) activity, suggesting a possible action of garlic on reducing the activation of low molecular weight carcinogens or nitrosamines, which are metabolized by this CYP (Le Bon et al., 2003). The authors concluded that consumption of garlic powder modulated the activities of carcinogen metabolizing enzymes. On the other hand, when rats were fed diets containing 0, 0.5, 2.0 and 5.0% of garlic powder for 8 weeks, beginning the diets with an intraperitoneal injection of diethylnitrosamine (DEN), concentrations of 2.0 and 5.0% of garlic powder significantly decreased the area and number of glutathione S-transferase positive foci (Park et al., 2002). Moreover, the diet also suppressed the CYP2E1 activity and protein levels in rats liver carcinogenesis initiated by DEN. Thus, the suppression of CYP2E1 by dietary garlic powder at levels of 2.0 and 5.0% could influence the preneoplastic foci formation and contribute to chemoprevention against the rat hepatocarcinogenesis.

RAW GARLIC

Raw garlic preparations can cause chemical burns on the skin, contact dermatitis, bronchial asthma (Harunobu Amagase, 2006), anemia, weight loss and failure to grow (Moriguchi et al., 2001). Hoshino et al. (2001) demonstrated that dehydrated raw garlic powder caused severe damage on the gastric mucosa. Moreover, Ma and Yin (2012) presented a case of anaphylaxis induced by raw garlic ingestion. In another study, the supplementation with raw garlic didn't protect LDL particles to Cu2+-mediated oxidation, while the LDL isolated from subjects receiving supplements with AGE were more resistant to oxidation (Munday et al., 1999). Moreover, Mahmoodi et al. (2006) evaluated the effect of raw garlic consumption on human blood biochemical factors in hyperlipidemic individuals, observing a significant reduction in the blood total cholesterol and triglycerides, while high density lipoprotein (HDL-C) significantly increased after the raw garlic consumption. Also, Harauma and Moriguchi (2006) demonstrated that raw garlic reduced systolic blood pressure in rats with spontaneous hypertension. Besides, raw garlic showed protection against isoproterenol-induced myocardial necrosis and associated oxidative stress (Banerjee et al., 2003). Recently, researchers demonstrated that raw garlic didn't cause effects on platelet inhibition, in contradiction to other preparations that exhibited antiaggregatory effects (Scharbert et al., 2007). Despite many controversies, further studies are needed to evaluate the real benefits and side effects of the raw garlic preparations.

CONCLUSION

The present mini-review summarizes the data on the effects of the different forms of the garlic. Safety is an obvious advantage of garlic, which has been confirmed by its multi-millennial consumption. However, the effects observed in this study are highly variable and deserve caution in using these popular formulations. Taking these facts into account, clinical trials aimed at revealing the efficacy of garlic preparations are of particular importance. Future research should standardize the dosage of garlic and type, i.e.,whether it should be taken fresh, cooked, or aged as well the benefits of the formulations and positive results of these trials may open a new era of the use of this traditional product.

REFERENCES

Aguilera P, ME Chánez-Cárdenas, A Ortiz-Plata, D León-Aparicio, D Barrera, M Espinoza-Rojo, J Villeda-Hernández, A Sánchez-García, Maldonado PD. (2010) Aged garlic extract delays the appearance of infarct area in a cerebral ischemia model, an effect likely conditioned by the cellular antioxidant systems. *Phytomedicine*. 17, 241-247.

Ali M, M Thomson, M Afzal. (2000) Garlic and onions: their effect on eicosanoid metabolism and its clinical relevance. *Prostaglandins Leukot Essent Fatty Acids*. 62, 55-73.

Allison GL, GM Lowe, K Rahman. (2006) Aged garlic extract and its constituents inhibit platelet aggregation through multiple mechanisms. *J. Nutr.* 136, 782S-788S.

Amagase H. (2006) Clarifying the real bioactive constituents of garlic. *J. Nutr.* 136, 716S-725S. Review.

Amorati R, GF Pedulli. (2008) Do garlic-derived allyl sulfides scavenge peroxyl radicals? *Org. Biomol. Chem.* 6, 1103-1107.

Ankri S, D Mirelman. (1999) Antimicrobial properties of allicin from garlic. *Microbes Infect*.1, 125-129. Review.

Anoush M, MA Eghbal, F Fathiazad, H Hamzeiy, NS Kouzehkonani. (2009) The protective effects of garlic extract against acetaminophen-induced oxidative stress and glutathione depletion. *Pak. J. Biol. Sci.* 12, 765-771.

Banerjee SK, S Sood, AK Dinda, TK Das, SK Maulik. (2003) Chronic oral administration of raw garlic protects against isoproterenol-induced myocardial necrosis in rat. *Comp. Biochem. Physiol. C. Toxicol. Pharmacol.* 136, 377-386.

Block E. (1992) The organosulfur chemistry of the genus Allium – implications for theorganic chemistry of sulfur. *Angew Chem. Int. Ed.* 31, 1135–1178.

Borek C. (2001) Antioxidant health effects of aged garlic extract. *J. Nutr.* 131, 1010S-5S. Review.

Borek C. (2006) Garlic reduces dementia and heart-disease risk. *J. Nutr.* 136, 810S-812S. Review.

Chandra-Kuntal K, Lee J, Singh SV. (2013) Critical role for reactive oxygen species in apoptosis induction and cell migration inhibition by diallyl trisulfide, a cancer chemopreventive component of garlic. *Breast Cancer Res Treat.*

Chung I, SH Kwon, ST Shim, KH Kyung. (2007) Synergistic antiyeast activity of garlic oil and allyl alcohol derived from alliin in garlic. *J. Food Sci.* 72, M437-440.

Cruz C, R Correa-Rotter, DJ Sánchez-González, R Hernández-Pando, PD Maldonado, CM Martínez-Martínez, ON Medina-Campos, E Tapia, D Aguilar, YI Chirino, J Pedraza-Chaverri. (2007) Renoprotective and antihypertensive effects of S-allylcysteine in 5/6 nephrectomized rats. *Am. J. Physiol. Renal. Physiol.* 293, F1691-698.

Eidi A, M Eidi, E. Esmaeili. (2006) Antidiabetic effect of garlic (Allium sativum L.) in normal and streptozotocin-induced diabetic rats. *Phytomedicine* 13, 624–629.

El-Demerdash FM, MI Yousef, NI Abou El-Naga. (2005) Biochemical study on the hypoglycemic e.ects of onion and garlic in alloxan-induced diabetic rats. *Food and Chemical Toxicology* 43, 57–63.

Gapter LA, OZ Yuin, KY Ng. (2008) S-Allylcysteine reduces breast tumor cell adhesion and invasion. *Biochem. Biophys. Res. Commun.* 367,446-451.

Gorinstein S, M Leontowicz, H Leontowicz, Z Jastrzebski, J Drzewiecki, J Namiesnik, Z Zachwieja, H Barton, Z Tashma, E Katrich, S Trakhtenberg. (2006) Dose-dependent influence of commercial garlic (Allium sativum) on rats fed cholesterol-containing diet. *J. Agric. Food Chem.* 54, 4022-4027.

Groppo FC, JC Ramacciato, RHL Motta, PM Ferraresi, A Sartoratto. (2007) Antimicrobial activity of garlic against oral streptococci. *Int. J. Dent. Hygiene.* 5, 109-115.

Harauma A, T Moriguchi. (2006) Aged garlic extract improves blood pressure in spontaneously hypertensive rats more safely than raw garlic. *J. Nutr.* 136, 769S-773S.

Hoshino T, N Kashimoto, S Kasuga. (2001) Effects of garlic preparations on the gastrointestinal mucosa. *J. Nutr.* 131, 1109S-1113S.

Ide N, BH Lau. (2001) Garlic compounds minimize intracellular oxidative stress and inhibit nuclear factor-kappa b activation. *J. Nutr.* 131, 1020S-1026S.

Larijani VN, Ahmadi N, Zeb I, Khan F, Flores F, Budoff M. (2013) Beneficial effects of aged garlic extract and coenzyme Q10 on vascular elasticity and endothelial function: the FAITH randomized clinical trial. *Nutrition.* 29, 71-5.

Lentini F. (2000) The role of ethnobotanics in scientific research. State of ethnobotanical knowledge in Sicily. *Fitoterap.* 1, S83-88.

Le Bon AM, MF Vernevaut, L Guenot, R Kahane, J Auger, I Arnault, T Haffner, MH Siess. (2003) Effects of garlic powders with varying alliin contents on hepatic drug metabolizing enzymes in rats. *J. Agric. Food Chem.* 51, 7617-7623.

Lee S, Joo H, Kim CT, Kim IH, Kim Y. (2012) High hydrostatic pressure extract of garlic increases the HDL cholesterol level via up-regulation of apolipoprotein A-I gene expression in rats fed a high-fat diet. *Lipids Health Dis.* 11, 77.

Ma S, Yin J. (2012) Anaphylaxis induced by ingestion of raw garlic. *Foodborne Pathog. Dis.* 9, 773-5.

Mahmoodi M, MR Islami, GR Asadi Karam, M Khaksari, A Sahebghadam Lotfi, MR Hajizadeh, MR Mirzaee. (2006) Study of the effects of raw garlic consumption on the level of lipids and other blood biochemical factors in hyperlipidemic individuals. *Pak. J. Pharm. Sci.* 19, 295-298.

Modem S, Dicarlo SE, Reddy TR. (2012) Fresh Garlic Extract Induces Growth Arrest and Morphological Differentiation of MCF7 Breast Cancer Cells. *Genes Cancer.* 3, 177-86.

Moriguchi T, N Takasugi, Y Itakura. (2001) The effects of aged garlic extract on lipid peroxidation and the deformability of erythrocytes. *J. Nutr.* 131, 1016S-1019S.

Munday JS, KA James, LM Fray, SW Kirkwood, KG Thompson. (1999) Daily supplementation with aged garlic extract, but not raw garlic, protects low density lipoprotein against in vitro oxidation. *Atherosclerosis.* 143, 399-404.

O'Gara EA, DJ Maslin, AM Nevill, DJ Hill. (2008) The effect of simulated gastric environments on the anti-Helicobacter activity of garlic oil. *J. Appl. Microbiol.* 104, 1324-1331.

Ohaeri OC, GI Adoga. (2006) Anticoagulant modulation of blood cells and platelet reactivity by garlic oil in experimental diabetes mellitus. *Biosci. Rep.* 26, 1-6.

Park KA, S Kweon, H Choi. (2002) Anticarcinogenic effect and modification of cytochrome P450 2E1 by dietary garlic powder in diethylnitrosamine-initiated rat hepatocarcinogenesis. *J. Biochem. Mol. Biol.* 35, 615-622.

Pedraza-Chaverri J, P Yam-Canul, YI Chirino, DJ Sánchez-González, CM Martínez-Martínez, C Cruz, ON Medina-Campos. (2008) Protective effects of garlic powder against potassium dichromate-induced oxidative stress and nephrotoxicity. *Food Chem. Toxicol.* 46, 619-627.

Polat ZA, A Vural, F Ozan, B Tepe, S Ozcelik, A Cetin. (2008) In vitro evaluation of the amoebicidal activity of garlic (Allium sativum) extract on Acanthamoeba castellanii and its cytotoxic potential on corneal cells. *J. Ocul. Pharmacol. Ther.* 24, 8-14.

Rahman K. (2003) Garlic and aging: new insights into an old remedy. *Ageing Res. Rev.* 2, 39-56. Review.

Razo-Rodríguez AC, YI Chirino, DJ Sánchez-González, CM Martínez-Martínez, C Cruz, J Pedraza-Chaverri. (2008) Garlic powder ameliorates cisplatin-induced nephrotoxicity and oxidative stress. *J. Med. Food.* 11, 582-586.

Ried K, Frank OR, Stocks NP. (2013) Aged garlic extract reduces blood pressure in hypertensives: a dose-response trial. *Eur. J. Clin. Nutr.* 67, 64-70.

Ross ZM, EA O'Gara, DJ Hill, HV Sleightholme, DJ Maslin. (2001) Antimicrobial properties of garlic oil against human enteric bacteria: evaluation of methodologies and comparisons with garlic oil sulfides and garlic powder. *Appl. Environ. Microbiol.* 67, 475-480.

Scharbert G, ML Kalb, M Duris, C Marschalek, SA Kozek-Langenecker. (2007) Garlic at dietary doses does not impair platelet function. *Anesth. Analg.* 105, 1214-1218.

Seo DY, Lee SR, Kim HK, Baek YH, Kwak YS, Ko TH, Kim N, Rhee BD, Ko KS, Park BJ, Han J. (2012) Independent beneficial effects of aged garlic extract intake with regular exercise on cardiovascular risk in postmenopausal women. *Nutr. Res. Pract.* 6, 226-31.

Singh DK, TD Porter TD. (2006) Inhibition of sterol 4alpha-methyl oxidase is the principal mechanism by which garlic decreases cholesterol synthesis. *J. Nutr.* 136, 759S-764S.

Sobenin IA, IV Andrianova, ON Demidova, T Gorchakova, AN Orekhov. (2008) Lipid-lowering effects of time-released garlic powder tablets in double-blinded placebo-controlled randomized study. *J. Atheroscler. Thromb.* 15, 334-338.

Sheen LY, CK Lii, SF Sheu, RH Meng, SJ Tsai. (1996) Effect of the active principle of garlic--diallyl sulfide--on cell viability, detoxification capability and the antioxidation system of primary rat hepatocytes. *Food Chem. Toxicol.* 34, 971-978.

Steiner M, W Li W. (2001) Aged garlic extract, a modulator of cardiovascular risk factors: a dose-finding study on the effects of AGE on platelet functions. *J. Nutr.* 131, 980S-984S.

Talalay P, P Talalay. (2001) The Importance of Using Scientific Principles in the Development of Medicinal Agents from Plants. *Acad. Medicine* 76, 238–247.

Tattelman E. (2005) Health Effects of Garlic. Am Fam Physician 72, 103-106. Review.

Thomson M, T Mustafa, M Ali. (2000) Thromboxane-B(2) levels in serum of rabbits receiving a single intravenous dose of aqueous extract of garlic and onion. *Prostaglandins Leukot. Essent. Fatty Acids.* 63, 217-221.

Yeh YY, L Liu. (2001) Cholesterol-lowering effect of garlic extracts and organosulfur compounds: human and animal studies. *J. Nutr.* 131, 989S-993S.

Yousuf S, Ahmad A, Khan A, Manzoor N, Khan LA. (2010) Effect of diallyldisulphide on an antioxidant enzyme system in Candida species. *Can. J. Microbiol.* 56, 816-21.

Youn HS, HJ Lim, HJ Lee, D Hwang, M Yang, R Jeon, JH Ryu. (2008) Garlic (Allium sativum) extract inhibits lipopolysaccharide-induced Toll-like receptor 4 dimerization. *Biosci. Biotechnol. Biochem.* 72, 368-375.

Yu CS, Huang AC, Lai KC, Huang YP, Lin MW, Yang JS, Chung JG. (2012) Diallyl trisulfide induces apoptosis in human primary colorectal cancer cells. *Oncol. Rep.* 28, 949-54.

Tsao SM, MC Yin. (2001) In-vitro antimicrobial activity of four diallyl sulphides occurring naturally in garlic and Chinese leek oils. *J. Med. Microbiol.* 50, 646-649.

Velliyagounder K, Ganeshnarayan K, Velusamy SK, Fine DH. (2012) In vitro efficacy of diallyl sulfides against the periodontopathogen Aggregatibacter actinomycetemcomitans. *Antimicrob. Agents Chemother.* 56, 2397-407.

Vlasova MA, Moshkovskii SA. (2006) Molecular interactions of acute phase serum amyloid A: possible involvement in carcinogenesis. *Biochemistry* (Mosc). 71, 1051-9.

In: Agricultural Research Updates. Volume 5
Editors: P. Gorawala and S. Mandhatri

ISBN: 978-1-62618-723-8
© 2013 Nova Science Publishers, Inc.

Chapter 7

VEGETABLE SOYBEAN (EDAMAME): PRODUCTION, PROCESSING, CONSUMPTION AND HEALTH BENEFITS

Bo Zhang,[1,] Cuiming Zheng,[2,†] Ailan Zeng,[3,‡] and Pengyin Chen[3,§]*

[1]Agricultural Research Station, Virginia State University, Virginia, US
[2]Syngenta Biotechnology, Research Triangle Park, North Carolina, US
[3]Department of Crop, Soil, and Environmental Sciences,
University of Arkansas, Fayetteville, Arkansas, US

ABSTRACT

Soybean is primarily grown as a commodity crop for its high protein and oil content in seed. In recent years, tremendous interests have been generated among consumers in soyfoods including tofu, natto, soymilk, soy-sprouts, soy-nuts, and edamame due to their nutritional value and health benefits. Edamame is a vegetable soybean harvested at full green pod stage, processed as in-pod or shelled seed product, marketed as fresh or frozen produce, and consumed in various forms of stews, salads, dips or salted snacks. As a rich source of protein, vitamins, calcium and isoflavones, edamame is a favorable food product for prevention of cardiovascular disease, mammary and prostate cancers, and osteoporosis. As a conventional (non-GMO) or organic crop, edamame requires specific varieties with desired seed size and quality attributes, specific cultural management practices for optimal production, and special processing procedures for quality product development. As consumers around the globe, particularly in the west, continue to incorporate edamame into their diets as healthy food products, edamame will rapidly

---

[*] Bo Zhang: Agricultural Research Station, Virginia State University, P.O. Box 9061, 203 M. T. Carter Building, VA 23806.
[†] Cuiming Zheng: Syngenta Biotechnology, Inc., 3054 Cornwallis Rd., Research Triangle Park, NC 27709.
[‡] Ailan Zeng: Department of Crop, Soil, and Environmental Sciences, 115 Plant Science Building, University of Arkansas, Fayetteville, AR 72701.
[§] Pengyin Chen: Department of Crop, Soil, and Environmental Sciences, 115 Plant Science Building, University of Arkansas, Fayetteville, AR 72701.

expand in the market place. For the producers, edamame as a cash crop has significant potential for substantial profits because of the high market value. This chapter provides an overview on edamame production including current production status and factors affecting production, processing procedures and conditions, consuming exploration and consumer acceptance, as well as the health benefits with an emphasis on nutrition.

Soybean [*Glycine max* (L.) Merr.] was introduced into the US in the early 1800's as a forage crop (Mease, 1804). Subsequently, it was transformed into a major commodity crop for oil and meal production. Soybean has also served as a relatively minor food crop for a long time. However, food-type soybean industry has spent 20 years growing from door to door sale to abundant supply throughout the US (Soyfoods Association of North America, 2011). The consumption of soyfood has increased not only because soybean is an important source of complete protein, but also because soyfood helps humans to reduce cardiovascular disease, osteoporosis, and cancer risks (Messina, 1999; Messina, 2009). Soyfood is typically classified into two categories based on seed size: small- and large-seeded. Tofu (soybean curd), edamame (vegetable soybeans), miso (fermented soup-base paste) and **soymilk** (soybeans soaked, ground fine, and strained) are made from large seed (> 20g/100 seeds), but for natto (fermented whole soybeans), soy sauce (tamari, shoyu, and teriyaki), and soy sprouts, small seeds (< 12g/100 seeds) are desirable (Zhang et al., 2011).

Vegetable soybean, known as *edamame* in Japan and *maodou* in China (Shurtleff and Lumpkin, 2001), is harvested at R6 to R7 growth stage when seed are still green (Fehr et al., 1971). Edamame is usually eaten as a snack in Japan and China after being boiled in salt water. It can also be stir-fried or boiled to add into stew or soup similar to sweet pea or lima bean (Mebrahtu and Devine, 2009). In addition to large seed size, edamame is desired to have high sugar content, smooth texture, nutty flavor, and lack of beany taste (Young et al., 2000; Mohamed and Rao, 2004; Mebrahtu and Devine, 2009). Recently, research on edamame has been focusing on breeding development including seed quality traits and diversity analysis, production improvement, consuming exploration, and marketing evaluation.

SEED QUALITY TRAITS

One of the main edamame breeding aims is to modify the nutritional content because seed compositions determine the product quality. Protein content of fresh immature edamame seeds is 83% of that of matured food-type soybean seeds. Protein content of fresh edamame seeds from 12 edamame varieties range from 333.2 to 386.0 g/kg with an average of 360.4 g/kg on dry weight basis (Rao et al., 2002). In contrast, protein content of mature large-seeded, food-type soybeans range from 390.0 to 487.0 g/kg with an average of 434.0 g/kg (Zhang et al., 2010). Likewise, oil content of edamame seed is lower than mature, large-seeded, food-type soybeans, with edamame seed ranging from 130.7 to 155.8 g/kg and mature seeds ranging from 156.0 to 217.0 g/kg in oil content on dry weight basis (Rao et al., 2002; Zhang et al., 2010). Total soluble sugar in edamame influences consumer acceptability and total sugar content is at its peak level at R6 stage (Masuda, 1991; Rao et al., 2002). Ten edamame genotypes were used for diallel analysis of sugar content (Mebrahtu and Devine, 2009). By testing total sugar and sucrose content of fresh seeds of F2 and F3 progenies, 'Verde', 'V81-1603', and PI 399055 were shown to be the best combiners for high total sugar content, while 'Kanrich', 'Pella', 'Verde', and 'V81-1603' exhibited good combining abilities for high sucrose content (Mebrahtu and Devine, 2009).

Amino acids are also important cellular components to affect edamame taste (Yanagisawa et al., 1997). Asparagine, alanine, and glytamic acid of immature seeds of two edamame varieties were tested at 20, 30, 40, and 50 days after flowering. The concentrations of all three amino acids were higher and edamame had more favorable taste at 30 to 40 days after flowering, than those at other periods (Yanagisawa et al., 1997). Soybean isoflavones consisting of 12 isomers help to protect humans against osteoporosis, breast and other common cancers, and cardiovascular diseases (Yoshiki et al., 1998; Kim et al., 2007). Embryo, cotyledon, seed coat and whole seed from four Korean edamame varieties were analyzed for 12 isoflavones. Embryo had highest average isoflavone concentration of 2147.8 ug g^{-1} followed by whole seed of 322.5 ug g^{-1} and cotyledon of 161.9 ug g^{-1}, whereas seed coat had lowest average isoflavone concentration of 33.6 ug g^{-1}. Malonylglycoside was the most abundant isoflavone form among the four isomer forms with an average of 60% in embryo to 83% in whole seed (Kim et al., 2007). Some studies suggested that herbicide application may increase isoflavones (Landini et al., 2002; Nelson et al., 2007). Hebicide lactofen was applied to increase isoflavones at postemergence of Envy, a MG III variety; however, lactofen did not influence isoflavone concentration in Envy seeds but reduced seed yield after lactofen was applied at R1 and R5 stages (Nelson et al., 2007).

PRODUCTION FACTORS

Edamame with high yield and superior quality is preferred by both farmers and consumers. Proper varieties offer potential for maximum yield and best quality, and thus the most economic return. Therefore, variety selection is the most important factor in edamame production. From 1995 to 1998 in Georgia, twelve Asian edamame accessions including six Japanese cultivars, four Japanese plant introductions, two Chinese edamame cultivars, and two elite US commodity soybean cultivars were evaluated for fresh bean yield and fresh seed composition, such as protein, oil, glucose, and phytate content (Rao et al., 2002). The fresh pod yields were significantly different among all genotypes, ranging from 14.6 Mg Ha^{-1} to 21.7 Mg Ha^{-1}, with 'Tambagura' being the highest and 'Ware' the lowest. Protein and oil content ranged from 333.2 to 386.0 g kg^{-1} and from 130.7 to 155.8 g kg^{-1}, respectively. The Japanese cultivars had slightly higher protein and lower oil content than the PIs and US cultivars. Glucose content varied from 60.3 to 74.0 g kg^{-1}, and 'Hutcheson' had the highest glucose content among all genotypes followed by 'Tamnagura' (71.4 g kg^{-1}). The mean phytate of all genotypes ranged from 10.8 ('Tambagura' and 'Haujaku') to 13.9 g kg^{-1} with a mean of 12.6 g kg^{-1} ('Guanyun Da Hei Dun') (Rao et al., 2002).

In 2003 and 2004 at North Dakota State University, five edamame varieties were evaluated for yield and quality traits such as seed size (Duppong and Hatterman-Valenti, 2005). The interactions of year by genotype, year, and genotype were significant on yield, pod/plant, and seed size. Total marketable yield of five genotypes ranged from 6484.2 to 11,364.5 kg/ha. Yield in 2003 was lower than that in 2004 due to low germination rate in 2003. 'Envy' produced smallest seeds, whereas 'Sayamusume' produced largest seeds among all five varieties in both years. 'Envy' and 'Sayamusume' also produced significantly higher percentage of two-bean pods than any other genotypes, but 'Butterbean', 'IA 1010', and 'IA 2062' produced higher percentage of three-bean pods than 'Envy' and 'Sayamusume'

(Duppong and Hatterman-Valenti, 2005). Four and 23 edamame varieties with maturity group from III to VII were evaluated in Mississippi in 2004 and 2005, respectively (Zhang and kyei-Boahen, 2007). The late-maturing varieties generally produced more pods per plant and more fresh marketable pods than early-maturing varieties. Total fresh pod yield ranged from 1615.2 to 21474.4 kg/ha in 2004, while the average fresh pod yield in 2005 was 29807.3 kg/ha when planted in April 18 and 20364.6 kg/ha when planted in May 10, respectively. 'Envy' had lowest yield among all varieties because of its small seeds and early maturity. 'Garden Soy 01', 'Garden Soy 21', 'Midori Giant', 'Mojo Green', and 'Moon Cake' had the highest fresh pod yield (Zhang and kyei-Boahen, 2007).

Production system was also studied to improve edamame yield. In Duppong and Hatterman-Valenti's study, irrigation was applied to edamame test plots three times (July 21, July 31, and Aug 16) in 2003 and five times (July 2, July 16, July 28, Aug 6, and Aug 18) in 2005 according to the NDSU Extension Service recommendations. No significant difference in agronomic performance was detected between irrigated and non-irrigated plots each season. Therefore, in some regions of North Dakota, rainfall may be adequate for edamame production (Duppong and Hatterman-Valenti, 2005).

PEST CONTROL IN EDAMAME PRODUCTION

Weed control with pre-emergence herbicides in edamame was evaluated by Pornprom et al. in 2010. Seven different single-dose herbicides and two tank mixes were applied on edamame cultivar No. 75 (from Thailand) field within one day of sowing. Application of alachlor 469 g a.i. ha[-1], clomazone 960 g a.i.ha[-1], metribuzin 525 g a.i. ha[-1], pendimethalin 1031.25 g a.i. ha[-1] + pendimethalin 928 g a.i. ha[-1] or tank mixed metribuzin 350 g a.i. ha[-1] + pendimethalin 928 g a.i. ha[-1] did not cause visual plant injury. Metribuzin 525 g a.i. ha[-1] was the most effective pre-emergence herbicide to control many weed species followed by pendimethalin 1031.25 g a.i. ha[-1] and tank mixed metribuzin 350 g a.i. ha[-1] + pendimethalin g a.i. ha[-1]. Metribuzin application also provided minimum visual injury on plants and had no effect on final yield. No phytotoxic effect from herbicide residues in the soil was observed on edamame biomass.

Insects also impose threat to production of edamame. McPherson et al. in 2008 reported that velvetbean caterpillar[*Anticarsia gemmatalis* (Hübner)] caused serious defoliation in some years. Other lepidopterans including soybean looper [*Pseudoplusia includens* (Walker)] and green clover-worm [*Hypena scabra* (F.)] were also observed in edamame fields. Stink bugs caused much seed damage, but cultivars had different level of stink bug resistance. Other commonly observed arthropods included threecornered alfalfa hoppers [*Spissistilus festinus* (Say)], grasshoppers (*Melanoplus* spp.). and potato leafhopper [*Empoasca fabae* (Harris)]. Edamame yield was not affected significantly by insects because most seed damage and defoliation happened when beans almost reached 80% pod cavity for harvest, but edamame quality and visual aesthetic was reduced and not accepted by consumers due to insect damage. Planting short maturing cultivars early would help to harvest edamame before insect peak occurrence in order to reduce insect damage. The insecticide (diflubenzuron + I-cyhalothrin) effectively reduced the number of arthropod and stink bugs, which helped to secure marketable quality of edamame.

EDAMAME PROCESSING

Edamame is mainly consumed frozen in pods as a snack or shelled beans in salad. Edamame products are blanched to stop trypsin inhibitor activity (TIA) and frozen to lengthen shelf life. The quality attributes were studied under different blanching and storage conditions (Mozzoni et al., 2009; Xu et al., 2012). Pods and shelled edamame beans had very similar quality attributes including TIA, color and texture after blanching, but sucrose content in beans with pods was higher than that in shelled beans because pods protected beans from leaching sucrose (Mozzoni et al., 2009). Blanching for 2-2.5 min reduced 80% of TIA and 98% of peroxidase activity, respectively (Mozzoni et al., 2009; Xu et al., 2012), indicating that enzyme activity was significantly prevented to maintain quality after 2-2.5 min blanching. The edamame color density was compared at different blanching time. Mozzoni et al. reported that 2 min blanching resulted in high density of green color, while Xu et al. found green color density peaked after 5 min blanching. The conclusions were slightly different in two studies because the color-measuring methods varied (i.e. different sample size, material, instrument, and blanching method). Canning of edamame was also studied by Mozzoni et al. in 2009. The higher concentration of $CaCl_2$ caused harder edamame texture with a linear relationship at the level of $CaCl_2$ below 3.5 g L^{-1}. Both $CaCl_2$ concentration and pH affected luminosity with quadratic effects ($P = 0.003$ and $P = 0.042$). The pH also influenced intensity of green color with a quadratic effect. Moreover, 90 s blanching only maintained 1% of the lipoxygenase activity to remove the beany flavor. The ideal canning condition for edamame was the brine containing 150 g L^{-1} NaCl and 2.9 g L^{-1} $CaCl_2$ at pH 6.5 to produce low acidity and retain pleasing color and texture (Mozzoni et al., 2009).

CONSUMPTION

Japan, China and Korea are the major producers and consumers of edamame. US edamame consumption is increasing because more consumers are recognizing the nutritional value and favorable flavor of edamame. For instance, although Japan demands 159,900 tons of edamame per year (Nguyen 1997), the edamame consumption in the US has quickly increased and approximately 14,877 tons per year was estimated by Johnson in 2000. Most edamame is marketed frozen because edamame harvest window is very short with only 7-10 days, and frozen edamame has longer shelf-life than fresh edamame (Montri et al., 2006). However, fresh edamame provides better quality in nutrition and flavor than frozen edamame, a large number of customers prefer fresh, in-pod edamame at farmers' markets in the US (Johnson, 2000; Montri et al., 2006).

The consumption method of edamame differs among different cultures. In China and Japan, edamame is mostly consumed as a cooked snack by squeezing or biting the beans out of pods. Chinese also use edamame seed as an ingredient in stir fry dishes such as stir fried chicken with edamame. Japanese prefer to use edamame as textural filling in desserts such as rice cake for its nutrition and light flavor (WiseGEEK.com, 2012). Japanese also prefer Zunda, sweetened edamame paste, served as a dessert alongside different types of rice cakes or ice cream. Zunda is also used as a flavoring in snacks and candy (Issasi, 2013). In Korea, edamame is usually mixed with rice and cooked together. In the US, shelled edamame beans are used in salads and pasta (Montri, 2006). New Edamame-based foods, such as edamame hummus, dips and freeze-dried nuts, are being developed as the US edamame market expands.

Consumers from different cultures have different preferences for edamame flavor. Desirable edamame varieties are characterized as having large seed size, bright green color, mild aroma, smooth, firm but tender texture, and sweet and nutty taste (Rodale Research Center, 1982; Carter and Shanmugasundaram 1993; Konovsky et al., 1994;). Additionally, consumers in Japan require edamame pods with light pubescence color, at least two beans per pod, sweet flavor and smooth texture (Konovsky et al., 1994). Consumers in South Asia and Africa dislike the beany flavor of edamame (Shanmugasundaram et al., 2004).

Edamame consumption is also affected by consumers' age, gender, income and ethnicity. Kelly et al. in 2005 reported that people of 41 years-old or older were more willing to buy edamame after being told about its health benefits; females paid more attention on edamame taste, whereas males more cared about pod appearance; customers with an income above $40, 000 were more likely to buy edamame; and Asian consumers bought edamame more often than non-Asian (Wszelaki et al., 2005). Additionally, consumer awareness of health benefits of edamame increase customer acceptance of edamame.Moreover, the study found that the customers with experience purchasing soyfoods are more likely to accept edamame (Kelly et al., 2005).

NUTRITION CONTENT

Edamame contains higher nutritional value than most other vegetable crops. It is beneficial for bone strength and cholesterol levels, and protects against cardiovascular disease, renal disease, bone resumption, certain forms of cancer, and menopausal symptoms (Messina, 2001).

Edamame contains higher protein content and lower oil content on a dry weight basis than grain-type soybeans. Edamame is rich in essential amino acids such as lysine and tryptophan, which do not exist in other grains. Soybean peptide which regulates leukocytes, granulocytes, lymphocytes and their subsets, has an important impact on their counterpart immune system markers and on the hormones involved in emotion of healthy volunteers (Yimit et al., 2012).

The average oil content of edamame was 29% lower than mature soybean seeds (Rao et. al., 2002). Significant variations in fatty acid compositions among edamame genotypes were observed in oleic (C18:1), linoleic (C18:2), and linolenic (C18:3) acids. These polyunsaturated fatty acids are essential fatty acids for normal growth of humans. They also reduce cholesterol levels in human blood, thereby reducing the risk of heart disease.

While glucose and total soluble sugar content vary among different edamame genotypes, the seed sugar content reaches the highest level in edamame at immature stage rather than mature, dry seed stage. Lipoxygenase is an anti-nutritional factor that develops off-flavor in soybean food and beverage, and reduces fatty acids in soy oil. Based on studies conducted by Mohamed in 1991 and 1992, lipoxygenase activity in edamame is lower than in grain-type soybean, and wide variations of lipoxygenase activity exist among edamame genotypes, making breeding and variety selection possible.

ISOFLAVONE CONTENT

Soy foods decrease the risk of atherosclerotic cardiovascular disease and improve endothelial function. Secondary metabolites such as phenolic compounds, isoflavones, saponins, and phytic acids are important healthy supplements. They provide pharmaceutical benefits to humans as antioxidants and anticarcinogens. Soybean isoflavones can be used as a postmenopausal hormone replacement therapy since they have the function of preventing osteoporosis, breast and ovarian cancer, and cardiovascular diseases.

There are 12 types of isoflavones which vary in concentration among soybean genotypes and across different environmental factors (Kim et al., 2007). Isoflavones accumulated in the order of malonylglycoside, glycoside, acetylglycoside, and aglycon, among which malonylglycoside was the most abundant form ranging from 66 to 79% of the total isoflavone content. Isoflavones concentrations accumulated the most in the embryo of fresh edamame (Kim et al., 2007), making the edamame products a healthy choice.

VITAMIN AND MINERAL CONTENT

Edamame contains vitamins and minerals that are an important nutritional source for human beings. Mohamed et al. (1991) studied protein, total phosphorus, available phosphorus, and minerals in edamame genotypes. It was indicated that there were significant differences for tested minerals, Ca, K, Fe, Mn, and Cu, among the genotypes. Mean Ca content was 2330.6 µg/g, and ranged from 1326.9 to 3262.9 µg/g. The PIs contained more Ca than US cultivars, and small-sized seeds contained significantly higher Ca than large-sized seeds. Potassium content varied among edamame genotypes, ranging from 85.9 to 178.4 µg/g with a mean of 116.8 µg/g. Potassium content in large-sized seed (120.0 µg/g) edamame was higher than small-sized seed (114.5 µg/g). Moderate variations in Fe content were observed ranging from 26.5 to 79.0 µg/g with a mean of 45.8 µg/g. In general, small-sized seeds contain more Fe (51.8 µg/g) than large-sized seeds (40.4 µg/g). Copper content in edamame ranged from 8.8 to 21.8 µg/g. The mean Cu for small-sized seeds (39.4 µg/g) was significantly higher than large-sized seeds (30.96 µg/g). Manganese content among genotypes varied more than the other selected vitamins and minerals. The mean of Mn was 16.3 µg/g and range from 3.9 to 68.4 µg/g. Large-sized seeds showed higher Mn concentration than did small-sized ones. In summary, the high content of Ca, Fe, and Mn in edamame make it a desirable vegetable option for consumers.

In summary, vegetable soybean has been very popular in Asian counties and is increasingly gaining acceptance in other parts of the world. As a vegetable, snack, or processed product, edamame provides significant nutrition value and health benefits.As the market grows and expands, there is a need for continued research and development focusing on variety development, production management, processing and marketing enhancements. Cultivars with high yield potential and improved quality attributes are needed for maximum productivity and profitability. It is important to select varieties with local adaptation, large seed, wide pod with 2 or 3 seed, dark green pod color with light colored pubescence, seed with high protein, high sucrose, high isoflavones, vitamins, and minerals, low stachyose, and lipoxygenase free. It is also valuable to have desirable sensory attributes such as proper texture, aroma flavor, and savory/nutty tastes.

Production of edamame is limited to small scale, mostly in China and Japan, due to the challenge of mechanical harvest. However, there are several larger scale commercial productions in the US and Canada where mechanized harvest is implemented. Varieties with different maturities are available in the US, but need to be improved for higher yield and better quality to compete with imported products. Challenges remain for production in terms of crop management, particularly in the area of weed and pest management, as limited labeled products are available for edamame as compared to other transgenic crops.

However, this crop can be profitable as varieties with high yield potential and wide maturity ranges are made available and market demand for high quality products increases. Processing of edamame, once harvested, is fairly straightforward either as in-pod or shelled products, fresh or frozen, and bulk or packaged. Secondary processed products such as freeze-dried veggie snacks, roasted nuts, dip, ice cream, and drinks are also possibilities for expanding the edamame market. Aa bright outlook for the edamame industry is ahead as consumers become more aware of the nutritional value and health benefits.

REFERENCES

Cater, T. E. and S. Shanmugasundaram. 1993. Edamame, the vegetable soybean. p.. 219–239. In: *Underutilized Crops: Pulses and Vegetables* (T. Howard, ed.) Chapman and Hill, London, UK.

Duppong, L. M. and H. Hatterman-Valenti. 2005. Yield and quality of vegetable soybean cultivars for production in North Dakota. *Hort Technology* 15:896–900.

Fehr, W., C. Caviness, D. Burmwood, and J. Pennington. 1971. Stage of development descriptions for soybeans. *Crop Sci.* 11:929-931.

Isassi, Yoko. "FoodStory | Meet + Eat + Discover » Zunda Mochi." FoodStory | Meet + Eat + Discover . http://ifoodstory.com/20110825/zunda-mochi/ (accessed January 4, 2013).

Landini, S., M. Y. Graham and T. L. Graham. 2002. Lactofen induces isoflavone accumulation and glyceollin elicitation competency in soybean. *Phytochemistry* 62:865–874.

Johnson, D. 2000. Edamame: Westerners develop a taste for Japanese soybean. *Eng. Technol. Sustainable World* 7:11-12.

Kelley, K. M. and E. S. Sanchez. 2005. Accessing and understanding consumer awareness of and potential demand for edamame. *Hortscience* 40: 1347-1353.

Kim, J. A., S. B. Hong, W. S. Jung, C. Y. Yu, K. H. Ma, J. G. Gwag, and I. M. Chung. 2003. Color, texture, nutrient contents, sensory values of vegetable soybeans [*Glycine max* (L.) Merrill] as affected by blanching. *Food Chem.* 83: 69-74.

Kim, J. A., S. B. Hong, W. S. Jung, C. Y. Yu, K. H. Ma, J. G. Gwag, and I. M. Chung. 2007. Comparison of isoflavones composition in seed, embryo, cotyledon and seed coat of cooked-with-rice and vegetable soybean (*Glycine max* L.) varieties. *Food Chem.* 102: 738-744.

Konovsky, J., T. A. Lumpkin and D. McClary. 1994. Edamame: The vegetable soybean. p. 173-181. In: A. D. O'Rourke (ed) Understanding the Japanese Food and Agrimarket: A Multifaceted Opportunity. Binghamton: Hayworth.

Masuda, R. 1991. Quality requirement and improvement of vegetable soybean, p. 92-102. In: S. Shanmugasundaram (ed.). *Vegetable soybean: Research needs for production and quality improvement*. Asian Veg. Res. Dev. Center, Taiwan.

McPherson, R. M., W. Johnson and E. G. Fonsah. 2008. Insect pests and yield potential of vegetable soybean (edamame) produced in Georgia. *J. Entomol. Sci.* 43:225-240.

Mease, J. 1804. Soy. In: A. F. M. Willich (ed.), *The Domestic Encyclopedia*. William Young Brich and Abraham Small, Philadelphia, PA. 5:12-13.

Mebrahtu, T. and T. E. Devine. 2008. Combining ability analysis for selected green pod yield components of vegetable soybean genotypes (*Glycine Max*). *N. Z. J. Crop Hort Sci.* 36:97-105.

Nelson, K., G. E. Rottinghaus and T. E. Nelson. 2007. Effect of lactofen application timing on yield and isoflavone concentration in soybean seed. *Agron. J.* 99:645-649.

Messina, M. J. 1999. Legumes and soybeans: Overview of their nutritional profiles and health effects. *Am. J. Clin. Nutr.* 70:439-450.

Messina, M. and V. Messina. 2000. Soyfoods, soybean isoflavones, and bone health: A brief overview. *J. Renal Nutr.* 10: 63-68.

Messina, M. 2001. An overview of the health effects of soyfoods and soybean isoflavones, p. 117-122. In: T. A. Lumpkin and S. Shanmugasundaram (comp.). *Proceedings, Second Int. Vegetable Soybean Conference*, Tacoma, WA. 10-12 Aug. 2001. Washington State Univ., Pullman, WA.

Messina, M. and A. H. Wu. 2009. Perspectives on the soy-breast cancer relation. *Am. J. Clinical Nutrition.* 89: 1673-1679.

Miles, C. A., T. A. Lumpkin and L. Zenz. 2000. Edamame. *Wash. State Coop. Ext. Bul.* PNW0525.

Mohamed, A. and M. S. Rao, 2004: Vegetable soybean as a functional food. p. 209-238. In: K. Liu (ed.), *Soybean functional foods and ingredients*. AOAC Champaign, Illinois Press.

Mohamed, A., T. Mebrahtu and M. Rangappa. 1991. Nutrient composition and anti-nutritional factors in selected vegetable soybean (*Glycine Max* [L.] Merr.) *Plant Food Hum. Nutr.* 41: 89-100.

Mohamed, A. and M. Rangappa. 1992. Screening soybean (grain and vegetable) genotypes for nutrients and anti-nutritional factors. *Plant Food Hum. Nutr.* 42:87-96.

Montri, D. N., K. M. Kelly and E. S. Sanchez. 2006. Consumer interest in fresh, in-shell edamame and acceptance of edamame-based patties. *Hortscience* 41: 1616-1622.

Mozzani, L., R. Morawicki and P. Chen. 2009. Canning of vegetable soybean: Procedures and quality evaluations. *Int. J. Food Sci. Technol.* 40:1125-1130.

Mozzani, L., P. Chen, R. Morawicki, N. Hettiarachchy, K. Brye, A. Mauromoustakos. 2009. Quality attributes of vegetable soybean as a function of boiling time and condition. *Int. J. Food Sci. Technol.* 44:2089-2099.

Nguyen, V. Q. 1997. Edamame (vegetable green soybean). *The new rural industries, a handbook for farmers and investors,* p. 1-7, 196-203. 10 Sept. 2003. www.rirdc.gov.au/pub/handbook/edamame.html.

Rao, M. S. S., A. S. Bhagsari and A. I. Mohamed. 2002. Fresh green seed yield and seed nutritional traits of vegetable soybean genotypes. *Crop Sci.* 42:1950-1958.

Rodale Research Center. 1982. Fresh green soybeans: Analysis of field performance and sensory qualities. Kutztown, PA: Rodale Press, 26 pp.

Shanmugasundaram, S. and M. R. Yan. 2004. Global expansion of high value vegetable soybean. p.915-920. In: *Proceedings of World Soybean Research Conference.* Shurtleff, W. and C. R. Lumpkin. 2001. Chronology of green vegetable soybeans and vegetable type soybeans. p. 97-103. In: T. A. Lumpkin and S. Shanmugasundaram (Compilers), *2nd Int. Vegetable Soybean Conf.*, Washington State Univ., Pullman.

Shurtleff, W. and A. Aoyagi. 2009. History of edamame, green vegetable soybeans, and vegetable-type soybeans (1275-2009): Extensively annotated bibliography and sourcebook, p.7-8. Lafayette, CA.

Wszelaki, A. L., J. F. Delwiche, S. D. Walker, R. E. Liggett, S. A. Miller, and M. D. Kleinhenz. 2005. Consumer liking and descriptive analysis of six varieties of organically grown edamame-type soybean. *Food Qual. Prefer.* 16: 651-658.

WiseGEEK, 2012. http://www.wisegeek.com/what-is-edamame.htm. Accessed on Novermber, 2012.

Xu, Y., E. Sismour, S. Pao, L. Rutto, and C. Grizzard. 2012. Textural and microbiological qualities of vegetable soybean (edamame) affected by blanching and storage conditions. *J. Food Process Technol.* 3:165. doi:10.4172/2157-7110.1000165.

Yimit, D., P. Hoxur, N. Amat, K. Uchikawa, and N. Yamaguchi. 2012. Effects of soybean peptide on immune function, brain function, and neurochemistry in healthy volunteers. *Nutrition* 28: 154-159.

Yanagisawa, Y., T. Akazawa, T. Abe, and T. Sasahara. 1997. Changes in free amino acid and kjeldahl N concentrations in seeds from vegetable-type and grain-type soybean cultivars during the cropping season. *J. Agric. Food Chem. 45:*1720–1724.

Yimit, D., P. Hoxur, A., Nurmuhammat. K. Uchikawa, N. Yamaguchi. 2012. Effects of soybean peptide on immune function, brain function, and neurochemistry in healthy volunteers. *Nutr.* 28: 154–159.

Yoshiki, Y., S. Kudou and K. Okubo. 1998. Relationship between chemical structure and biological activities of triterpenoid saponin from soybean. *Biosci., Biotechnol., Biochem.* 62:2291–2299.

Young, G., T. Mebrahtu and J. Johnson. 2000. Acceptability of green soybeans as a vegetable entity. *Plant Foods Human. Nutr.* 55:323-333.

Zhang, B., L. Jaureguy, L. Florez-Palacios, and P. Chen. 2011. Food-grade soybean: Consumption, nutrition, and health. In: Bean: consumption, nutrition, and health. E. Popescu and I. Golubev (ed.) *Beans: consumption, nutrition, and health.* Nova Science Publishers, Inc., Hauppauge, NY.

Zhang, B., P. Chen, S. L. Florez-Palacios, A. Shi, A. Hou, and T. Ishibashi. 2010. Seed quality attributes of food-grade soybeans from the US and Asia. *Euphytica* 173: 387-396.

Zhang, L. and S. Kyei-Boahen. Growth and yield of vegetable soybean (edamame) in Mississippi. *HortTechnology.* 17:26-31.

In: Agricultural Research Updates. Volume 5
Editors: P. Gorawala and S. Mandhatri

ISBN: 978-1-62618-723-8
© 2013 Nova Science Publishers, Inc.

Chapter 8

NON-SIGNIFICANT INTERACTIONS BETWEEN TREATMENTS AND A SUGGESTED STATISTICAL APPROACH FOR DEALING WITH THESE STATUSES

Zakaria M. Sawan

Cotton Research Institute, Agricultural Research Center,
Ministry of Agriculture and Land Reclamation, Giza, Egypt

ABSTRACT

A field experiment was conducted to study the effect of nitrogen (N) fertilizer and foliar application of potassium (K) and Mepiquat Chloride (MC) on yield of cotton. Seed cotton yield per plant and seed cotton and lint yield per hectare; have been increased due to the higher N rate and use of foliar application of K and MC. No significant interactions were found among the variables in the present study (N, K and MC) with respect to characters under investigation. Generally, interactions indicated that, the favorable effects ascribed to the application of N; spraying cotton plants with K combined with MC on cotton productivity, were more obvious by applying N at 143 kg per hectare, and combined with spraying cotton plants with K at 957 g per hectare and also with MC at 48 + 24 g active ingredient per hectare. Sensible increases were found in seed cotton yield per hectare (about 40%) as a result of applying the same combination.

However, this interaction did not reach the level of significance, so, statistical approach for dealing with the non-significant interactions between treatments, depending on the Least Significant Difference values has been suggested, to provide an opportunity to disclosure of the interaction effects regardless of their insignificance. As a matter of fact the original formula used in calculating the significance of interactions suffers a possible shortage, which can be eliminated through applying the new suggested formula (Sawan, 2011).

Keywords: Cotton Yield; Mepiquat Chloride; Nitrogen; Non-Significant Interactions; Potassium

1. INTRODUCTION

Managing the balance of vegetative and reproductive growth is the essence of managing a cotton crop. It is well known from numerous fertilizer experiments that the yield of field crop has been dependent strongly on the supply of mineral nutrients (Gormus, 2002; Ansari and Mahey, 2003; Pervez et al., 2004). Excess of vegetative growth, poor bud development, shedding of fruiting forms, and growth imbalance between the source and sink are responsible for the unpredictable behavior of the crop. Several approaches have tried-out to break this yield plateau, among them the application of plant growth regulators (PGR's), particularly Mepiquat Chloride (MC) that has received greater attention recent years (Kumar et al., 2004; Nuti et al., 2004).

The objective of this study was to evaluate the effects of N fertilization rate, foliar K application, and MC application on the yield of cotton with the aim to identify production treatments that may improve the yield. Also, we suggested a statistical approach for dealing with the non-significant interactions between treatments depending on the Least Significant Difference values, regardless of statistical insignificance (Sawan, 2011).

2. MATERIALS AND METHODS

A Field experiment was conducted at the Agricultural Research Center, Ministry of Agriculture in Giza (30°N, 31°: 28'E and 19 m altitude), Egypt using the cotton cultivar 'Giza 86' (*Gossypium barbadense* L.) in I and II seasons. The soil texture in both seasons was a clay loam, with an alluvial substratum, (pH = 8.10, 44.75% clay, 27.40% silt, 20.00% fine sand, 3.00% coarse sand, 2.85% calcium carbonate and 1.85% organic matter). Each experiment included 16 treatment combinations of: (i) two N rates (95 and 143 kg N per hectare), which were applied as ammonium nitrate (NH_4NO_3, 33.5% N) at two equal doses, 6 and 8 weeks after planting. Each application (in the form of pinches beside each hill) was followed immediately by irrigation. (ii) four K rates (0, 319, 638 and 957 g K per hectare) were applied as potassium sulfate (K_2SO_4, '40% K') as a foliar spray, 70 and 95 days after planting (during square initiation and boll development stage). The solution volume applied was 960 L per hectare. (iii) two rates from the PGR, 1,1-dimethylpiperidinium chloride (Mepiquat Chloride 'MC' or 'Pix') were foliar applied (75 days after planting at 0 or 48 g active ingredient per hectare, 90 days after planting at 0 and 24 g active ingredient per hectare) where the solution volume applied was also 960 L per hectare. The K and MC were applied to the leaves with uniform coverage using a knapsack sprayer. The pressure used was 0.4 kg per cm^2, resulting in a nozzle output of 1.43 L per min. The application was carried out between 9.0 and 11.0 h (Sawan, 2011).

A randomized complete block design with four replications was used for both experiments. Seeds were planted on 3 April, in season I and 20 April, in season II. Plot size was 1.95 × 4 m including three ridges (beds) (after the precaution of border effect was taken into consideration). Hills were spaced 25 cm apart on one side of the ridge, with seedlings thinned to two plants hill^{-1} six weeks after planting. This provided a plant density of 123,000 plants per hectare. The total amount of irrigation applied during the growing season (surface irrigation) was about 6,000-m^3 per hectare. The first irrigation was applied three weeks after planting, with the second three weeks later. Thereafter, plots were irrigated every two weeks until the end of the season (October 11, in season I and October 17, in season II, respectively), for a total of nine irrigations. On the basis of soil test results, phosphorus (P) fertilizer was applied at the rate of 24 kg P per hectare as calcium super phosphate during land preparation. The K fertilizer was applied at the rate of 47 kg K per hectare as potassium sulfate before the first irrigation (the recommended level for semi-fertile soil). Fertilization (P and K), along with pest and weed management was carried out during the growing season; according to the local practice performed at the experimental station (Sawan, 2011).

In both seasons, ten plants were randomly taken from the center ridge of each plot to determine the seed cotton yield in g per plant. First hand picking was made on 20 and 26 September and final picking on 11 and 17 October in season I, and season II, respectively. Total seed cotton yield of each plot (including ten plant sub samples) was ginned to determine seed cotton and lint yield (kg per hectare) (Sawan, 2011).

Results were analyzed as a factorial experiment in a randomized complete block design for the studied characters each season and the combined statistical analysis for the two seasons, following the procedure outlined by Snedecor and Cochran (1980). The Least Significant Difference (L.S.D.) test method, at 5% level of significance was used to verify the significance of differences among treatment means and the interactions to determine the optimum combination of N, K and MC (Sawan, 2011).

3. RESULTS AND DISCUSSION

Results from the analysis of variance for yield (combined data of the two seasons) are presented in Table 1 (Sawan, 2011).

Table 1. Mean squares for combined analysis of variance for yield in cotton during season I and season II

Source	d.f.	Seed cotton yield (g per plant)	Seed cotton yield (kg per hectare)	Lint yield (kg per hectare)
Year	1	147.21**	1415571.4**	332917.8**
Replicates within years	6	40.27*	404859.0*	50458.4*
Treatments	15	75.94**	714189.8**	83868.9**
Nitrogen (N)	1	456.74**	4325402.3**	500162.5**
Potassium (K)	3	132.53**	1223590.9**	145491.8**
Mepiquat Chloride (MC)	1	261.15**	2504937.5**	294768.0**
N × K	3	3.47	31778.5	3934.8
N × MC	1	0.17	1463.4	298.6
K × MC	3	4.19	36432.4	4632.6
N × K ×MC	3	0.18	1879.3	209.1
Treatments × Year	15	2.50	24239.8	3070.9
Error	90	14.36	135377.4	16752.8
SD		3.79	367.9	129.4
CV %		12.04	12.0	12.0

* Significant at $P = 0.05$.
** Significant at $P = 0.01$.
(Sawan, 2011).

3.1. Effects of Main Treatments on Yield

Seed cotton yield per plant, as well as seed cotton and lint yield per hectare, were increased by as much as 12.8, 12.8, and 12.3 %, respectively, when the nitrogen rate was increased (Table 2) (Sawan, 2011).

There were both increased boll numbers and boll weight, which was attributed to the fact that N is an important nutrient for control of new growth and preventing abscission of squares and bolls and is also essential for photosynthetic activity (McConnell and Mozaffari, 2004; Wiatrak et al., 2006).

When K was applied at all three K rates (319, 638 and 957 g K per hectare), seed cotton yield plant[-1] and seed cotton and lint yield ha[-1] were also increased (Sawan, 2011). These increases could be attributed to the favorable effects of K on yield components, i.e. number of opened bolls per plant, and boll weight, leading consequently to higher cotton yield (Pettigrew et al., 2005; Sharma and Sundar, 2007).

Mepiquat Chloride, significantly increased seed cotton yield per plant, as well as seed cotton and lint yield per hectare (by 9.5, 9.6, and 9.3%, respectively), compared to the untreated control (Sawan, 2011).

These results may be attributed to the promoting effect of this substance that has beneficial and supplemental affects leading to yield enhancement (boll retention and boll weight) (Sharma and Sundar, 2007).

**Table 2. Effect of N-rate and foliar application of K and MC
on yield in cotton combined over two seasons I and II**

Treatment	Seed cotton yield (g per plant)	Seed cotton yield (kg per hectare)	Lint yield (kg per hectare)
N rate (kg per hectare)			
95	29.58[b]	2882.3[b]	1020.0[b]
143	33.36[a]	3250.0[a]	1145.0[a]
LSD (0.05)	1.33	128.9	45.4
K rate (g per hectare)			
0	28.61[b]	2792.5[b]	988.2[b]
319	31.51[a]	3068.6[a]	1083.4[a]
638	32.51[a]	3163.0[a]	1115.2[a]
957	33.25[a]	3240.7[a]	1143.1[a]
LSD (0.05)	1.88	182.3	64.1
MC rate (g per hectare)			
0	30.04[b]	2926.3[b]	1034.5[b]
48 + 24	32.90[a]	3206.1[a]	1130.5[a]
LSD (0.05)	1.33	128.9	45.4
SD	3.79	367.9	129.4
CV %	12.04	12.0	12.0

Values followed by the same letter in a column are not significantly different from each other at $P = 0.05$.
(Sawan, 2011).

3.2. Effects of Interactions between Treatments on Yield

No significant interactions were found among the variables in the present study (N rates, K rates and MC) with respect to the characters under investigation. Generally, interactions indicated that, the favorable effects accompanied the application of N; spraying cotton plants with K combined with MC on cotton productivity, was more obvious by applying N at 143 kg per hectare, and combined with spraying cotton plants with K at 957 g per hectare and also with MC at 48 + 24 g active ingredient per hectare. Regarding the non-significant interaction effects, sensible increases were found in seed cotton yield per hectare (about 40%) as a result of applying the same combination (Sawan, 2011).

In this experiment there are sensible differences between the interactions, i.e. the first order (Tables 3-5), and the second order (Table 6).

Table 3. Effect of interaction between N rate and foliar application of K on yield combined over two seasons I and II

Character	Seed cotton yield (g per plant)		Seed cotton yield (kg per hectare)		Lint yield (kg per hectare)	
K rate	N rate (kg per hectare)					
(g per hectare)	95	143	95	143	95	143
0	27.04^d	30.18^c	2639.2^d	2945.8^c	936.0^d	1040.3^c
319	29.73^c	33.28^{ab}	2896.6^c	3240.5^{ab}	1025.3^c	1141.5^{ab}
638	30.16^c	34.86^a	2935.5^c	3390.4^a	1037.2^c	1193.3^a
957	31.38^{bc}	35.11^a	3058.0^{bc}	3423.3^a	1081.4^{bc}	1204.7^a
LSD (0.05) †	2.66		257.8		90.7	

Values followed by the same letter in columns under every character head are not significantly different from each other at $P = 0.05$.

† LSD, Least significant difference.

(Sawan, 2011).

Table 4. Effect of interaction between N rate and foliar application of MC on yield combined over two seasons I and II

Character	Seed cotton yield (g per plant)		Seed cotton yield (kg per hectare)		Lint yield (kg per hectare)	
N rate	MC rate (g per hectare)					
(kg per hectare)	0	48 + 24	0	48 + 24	0	48 + 24
95	28.11^c	31.04^b	2739.1^c	3025.6^b	970.4^c	1069.5^b
143	31.96^b	34.75^a	3113.5^b	3386.5^a	1098.5^b	1191.4^a
LSD (0.05)†	1.88		182.3		64.1	

Values followed by the same letter in columns under every character head are not significantly different from each other at $P = 0.05$.

† LSD, Least significant difference.

(Sawan, 2011).

Table 5. Effect of interaction between K rate and foliar application of MC on yield combined over two seasons I and II

Character	Seed cotton yield (g per plant)		Seed cotton yield (kg per hectare)		Lint yield (kg per hectare)	
K rate (g per hectare)	MC rate (g per hectare)					
	0	48 + 24	0	48 + 24	0	48 + 24
0	27.22^c	29.99^b	2655.0^c	2930.0^b	941.1^c	1035.3^b
319	29.66^{bc}	33.35^a	2891.3^{bc}	3245.8^a	1022.0^{bc}	1144.9^a
638	31.00^b	34.03^a	3014.1^b	3311.8^a	1064.2^b	1166.3^a
957	32.28^{ab}	34.21^a	3144.7^{ab}	3336.6^a	1110.7^{ab}	1175.5^a
LSD (0.05) †	2.66		257.8		90.7	

Values followed by the same letter in columns under every character head are not significantly different from each other at $P = 0.05$.

† LSD, Least significant difference.

(Sawan, 2011).

Table 6. Effect of interactions between N rate, foliar application of K and MC on yield in cotton combined over two seasons I and II

Treatment			Seed cotton yield (g per plant)	Seed cotton yield (kg per hectare)	Lint yield (kg per hectare)
N rate (kg per hectare)	K rate (g per hectare)	MC rate (g per hectare)			
95	0	0	25.54^e	2490.4^e	884.4^e
		48 + 24	27.85^{de}	2716.3^{de}	963.2^{de}
	319	0	28.71^{de}	2793.6^{de}	987.6^{de}
		48 + 24	30.36^{cd}	2956.1^{cd}	1046.7^{cd}
	638	0	28.54^{de}	2788.0^{de}	987.6^{de}
		48 + 24	31.62^{bcd}	3077.0^{bcd}	1087.4^{bcd}
	957	0	31.62^{bcd}	3077.4^{bcd}	1086.7^{bcd}
		48 + 24	32.40^{bc}	3160.0^{bc}	1116.2^{bc}
143	0	0	28.91^{cd}	2819.7^{cd}	997.8^{cd}
		48 + 24	31.48^{bcd}	3066.3^{bcd}	1080.8^{bcd}
	319	0	33.28^{ab}	3234.7^{ab}	1140.8^{ab}
		48 + 24	34.20^{ab}	3333.4^{ab}	1174.7^{ab}
	638	0	31.45^{bc}	3072.0^{bc}	1082.9^{bc}
		48 + 24	35.08^{ab}	3414.7^{ab}	1202.3^{ab}
	957	0	36.44^a	3546.2^a	1245.8^a
		48 + 24	36.03^a	3513.2^a	1234.8^a
LSD (0.05) †			3.76	364.6	128.3

Means followed by the same letter in a column are not significantly different from each other at $P = 0.05$.

† LSD, Least significant difference.

(Sawan, 2011).

However, these interactions did not reach the level of significance, so, suggested statistical approach for dealing with the non-significant interactions between treatments, depending on the Least Significant Difference values to verify the significant between treatment combinations regardless of the non-significance of the interaction effects from the

ANOVA, to reach a balance between experience and level of statistics as shown in Tables 3-6. It is quite possible that the experimental error could mask the pronounced effects of the interactions (Sawan, 2011).

In this manner, we found from the results that, if there were no significant differences existed between the different levels of any main factor (N, K or MC), in such case if the Least Significant Difference was calculated, the significance would not be existed. On the other hand, if the significance of the interactions between the main factors (first and second order interactions) did not existed, the estimation of Least Significant Difference of the interactions between the main factors, could give significant result (Sawan, 2011).

Thus, it could be said that the formula used in calculating the significance of interactions suffers a possible shortage.

We think that, it could be useful to modify or add some additions to the original formula used for calculating F values of interactions (Sawan, 2011):

F = Mean square for interaction / Mean square for error

In this connection, we could suggest that when calculating the significance of interactions we could calculate it as follow:

F = Mean square for interaction $\times$ n / Root of mean square for error (suggested formula)

Where n = number of main factors in the interaction.

We strongly believe that the use of the suggested formula, would secure the disclosure of the significant effects of the interactions regardless of the high experimental error (Sawan, 2011).

REFERENCES

Ansari, M.S. and Mahey, R.K. (2003) Growth and yield of cotton species as affected by sowing dates and nitrogen levels. *Journal of Research, Punjab Agricultural University*, 40, 8-11.

Gormus, O. (2002). Effects of rate and time of potassium application on cotton yield and quality in Turkey. *Journal of Agronomy and Crop Science*, 188, 382-388.

Kumar, K.A.K., Patil, B.C. and Chetti, M.B. (2004) Effect of plant growth regulators on biophysical, biochemical parameters and yield of hybrid cotton. *Karnataka Journal of Agricultural Science*, 16, 591-594.

McConnell, J.S. and Mozaffari, M. (2004) Yield, petiole nitrate, and node development responses of cotton to early season nitrogen fertilization. *Journal of Plant Nutrition*, 27, 1183-1197.

Nuti, R.C., Witten, T.K., Jost, P.H. and Cothren, J.T. (2000) Comparisons of Pix Plus and additional foliar *Bacillus cereus* in cotton. In *Proceedings Beltwide Cotton Production Research Conference*, San Antonio, TX, USA, January 4-8, Memphis, USA; Natl. Cotton Council pp. 684-687.

Pervez, H., Ashraf, M. and Makhdum, M.I. (2004) Influence of potassium rates and sources on seed cotton yield and yield components of some elite cotton cultivars. *Journal of Plant Nutrition*, 27, 1295-1317.

Pettigrew, W.T., Meredith, W.R.Jr. and Young, L.D. (2005) Potassium fertilization effects on cotton lint yield, yield components, and reniform nematode populations. *Agronomy Journal*, 97, 1245-1251.

Ram Prakash, Mangal Prasad and Pachauri, D.K. 2001. Effect of nitrogen, chlormequat chloride and FYM on growth yield and quality of cotton (*Gossypium hirsutum* L.). *Annals of Agricultural Research*, 22, 107-110.

Sawan, Z. M. (2011) A suggested statistical approach for dealing with the non-significant interactions between treatments. *Natural Science*, 3, 365-370.

Sharma, S.K. and Sundar, S. (2007) Yield, yield attributes and quality of cotton as influenced by foliar application of potassium. *Journal of Cotton Research and Development*, 21, 51-54.

Snedecor, G.W. and Cochran, W.G. (1980) *Statistical Methods*. 7th Ed. Iowa State University Press. Ames, Iowa, USA.

Wiatrak, P.J., Wright, D.L. and Marois, J.J. (2006) Development and yields of cotton under two tillage systems and nitrogen application following white lupine grain crop. *Journal Cotton Science*, 10, 1-8.

In: Agricultural Research Updates. Volume 5
Editors: P. Gorawala and S. Mandhatri

ISBN: 978-1-62618-723-8
© 2013 Nova Science Publishers, Inc.

Chapter 9

RAPID METHODS FOR IMPROVING NUTRITIONAL QUALITY OF HYDROPONIC LEAFY VEGETABLES BEFORE HARVEST

Wenke Liu,[*] Lianfeng Du and Qichang Yang

Environment and Sustainable Development in Agriculture, Chinese Academy of Agricultural Sciences, Key Lab. for Energy Saving and Waste Disposal of Protected Agriculture, Ministry of Agriculture, Beijing, China

ABSTRACT

Hydroponics is an important cultivation method to produce leafy vegetables in protected facilities with many of advantages compared with soil culture. However, some nutritional quality problems often occur for hydroponic leafy vegetables cultivated under cover due to heavy nitrogen fertilizer supply and low light intensity, e.g. high level of nitrate accumulation, low content of vitamin C and soluble sugar and so on. In this chapter, two established short-term methods, i.e. nitrogen deprivation method and short-term continuous lighting method, to improve nutritional quality of hydroponic leafy vegetables before harvest were summarized. Additionally, the regulation strategy for promoting nutritional quality of hydroponic leafy vegetables before harvest and its advantages were discussed.

Keywords: Nutritional quality; nitrate; vitamin C; hydroponics; pre-harvest regulation

INTRODUCTION

Hydroponics is a kind of well established method to produce horticultural crops supplying mineral nutrients via nutrient solution under cover. Hydroponics has many of

---

[*] Corresponding author: Wenke Liu. Environment and Sustainable Development in Agriculture, Chinese Academy of Agricultural Sciences, Key Lab. for Energy Saving and Waste Disposal of Protected Agriculture, Ministry of Agriculture, Beijing 100081. E-mail: liuwke@163.com.

advantages compared with soil culture. To sum up, four advantages of hydroponics are included. Firstly, the nutrients in nutrient solution are adjustable. So grower can apply different nutrient solutions for various growth stages of horticultural crops to maximize the yield and nutrient utilization efficiency. Second, hydroponics can avoid the occurrence of soil continuous cropping obstacles, and decrease the risk of diseases. Third, hydroponics can increase the culture area of protected horticulture due to no demand on soil quality. Therefore, hydroponics can be carried out on some lands without soil or with poor soil quality. Additionally, hydroponics-cultivation vegetables are uniform, surface clean and high yield. In past decades, hydroponics technology had been improved and applied worldwide in protected facilities, particularly in Netherland, China and Japan etc. More importantly, it is well known that hydroponics is more suitable for culture of fruit and leafy vegetables. However, some nutritional quality problems often occur for hydroponic leafy vegetables cultivated under cover due to heavy nitrogen fertilizer supply and low light intensity, e.g. high level of nitrate accumulation, low vitamin C and soluble sugar contents and so on.

Vegetables, in particular leafy vegetables, are dominant sources of nitrate intake through dietary pathway for human due to high level nitrate accumulation and large consumption. It was estimated that approximately 80% nitrate intake of human body was from dietary vegetable consumption, and excessive intake of nitrate will pose a potential hazard to public health, especially for infants (Eichholzer and Gutzwiller 1998). The toxic effects of nitrate are due to its endogenous conversion to nitrite, which is implicated in the occurrence of methaemoglobinaemia, gastric cancer and many other diseases (Santamaria, 2006). Today, off-season vegetable production in winter and spring, are usually cultivated under protected conditions worldwide. Many investigations around the world have found that nitrate accumulation in vegetables was serious (Zhou et al., 2000; Zhong et al., 2002; Chung et al., 2003; Sušin et al., 2006).

As a result, nitrate would be accumulated in vegetables when leafy vegetables were cultured under protected conditions for the weaker inner light intensity. Therefore, to develop efficient measures to decrease nitrate content in vegetables is a hot research issue worldwide in the past more than twenty years. Nowadays, based on our knowledge, it has realized that nitrate content in vegetables cultivated soilless can be successfully controlled through nutrient solution regulation and environmental factor control.

Therefore, developing measures to reduce nitrate concentration in vegetables are effective method to lower dietary nitrate instead of changing consumption, as well as working out the compulsive policy. Today, many countries had issued the limits for nitrate concentration of vegetables (Table 1). Meantime, low vitamin C content and soluble sugar content often synchronize with high level of nitrate accumulation in leafy vegetables.

Vitamin C and soluble sugar are health-beneficial substances that are rich in fresh vegetables. Vitamin C is soluble antioxidant substance to prevent human from some diseases, and soluble sugar is carbohydrate that can provide energy for human.

Therefore, to improve vitamin C and soluble sugar contents and to decrease nitrate accumulation in hydroponic leafy vegetables are urgent issue for sustainable development of hydroponics application in protected horticulture production. In this chapter, two established short-term methods, i.e. nitrogen deprivation method and short-term continuous light method, to improve nutritional quality of hydroponic leafy vegetables before harvest were summarized. Additionally, the regulation strategy for promoting nutritional quality of leafy vegetables before harvest and its advantages were discussed.

Table 1. Guide and maximum tolerated nitrate concentrations of vegetables
(mg NO$_3$ kg^{-1} in fresh weight)

Species	Germany (guide)	Netherland (Maximum)	Switerland (Guide)	Austrilia (Maximum)	Russia (Maximum)	EC (Maximum)
Lettuce	3000	3000(S) 4500(W)	3500	3000(S) 4000(W)	2000(O) 3000(G)	3500(4-10) 4500(11-3) 2500(O,5-8)
Spinach	2000	3500(S) 4500(W) 2500(1995)	3500	2000(<7) 3000(>7)	2000(O) 3000(G)	2500(4-10) 3000(11-3) 2000(P)
Red beet	3000	4000(4-6) 3500(7-3)	3000	3500(S) 4500(W)		
Radish	3000			3500(S) 4500(W)		
Endive cabbage		3000(S)	875	2500 1500	900(S) 500(W)	
Carrot				1500	400(S) 250(W)	

Note: S:summer. W:winter. O:outdoor. G:greenhouse. P:processed product (preserved/frozen).

<7: harvest by the end of June. >7: harvest from July. 1995: from 1995. 4-10:1 April to 31 October. 11-3:1 November to 31 March. 5-8:1 May to 31 August. Data from Sohn and Yoneyama (1996) and MAFF UK (1999).

1. SHORT-TERM TREATMENT METHOD BEFORE HARVEST

Many literatures had revealed that the causes of heavy accumulation of nitrate as well as low level of vitamin C and soluble sugar in hydroponic leafy vegetables are attributed to overuse of nitrate nitrogen fertilizer and low irradiation under cover. So, to reduce nitrate nitrogen supply and to elevate irradiation level is helpful to reduce nitrate level and improve vitamin C content in hydroponic leafy vegetables.

Nowadays, two kinds of methods have been developed. The first is the entire-process control method, i.e. controlling nitrate nitrogen supply or increasing irradiation level in whole cultivation process. The second is the short-term treatment method, i.e. controlling nitrate nitrogen supply or increasing irradiation level in several days before harvest.

Compared with entire-process control method, short-term treatment method before harvest shows a lot of advantages. This method can save a large of resources, energy and treatment time, labor, while regulation efficiency is greatly improved in terms of yield and nutrition quality.

Also, long-term nitrogen deprivation treatment will result in yield reduction. Short-term treatment method is characterized of short-term treatment, high-efficacy and resources-saving. Up to now, two established short-term methods, i.e. nitrogen deprivation method and short-term continuous lighting method, have been developed to reduce nitrate content, and to increase vitamin C and soluble sugar contents.

1.1. Nitrogen Deprivation Method

Nitrogen deprivation method is referred to stop supplying nitrogen fertilizer to hydroponic leafy vegetables during several days before harvest. After treatment, nitrate content is decreased by 30% to 90%. Usually, tap water or water solutions with osmotic ions were used to replace nutrient solution. Many studies had evidenced that this method is efficient in lowering nitrate content in hydroponic leafy vegetables. However, this method is high efficient only under the condition with enough light irradiation because nitrate in tissue of leafy vegetables can just be rapidly reduced under suitable light intensity (Liu et al., 2012a).

Two glasshouse experiments were conducted to investigate the efficiencies of various hydroponic solutions and their concentration effects in reducing nitrate concentration of lettuce after short-term treatment (Liu et al., 2011). The first experiment showed that hydroponic solutions of 0.1 mM KCl, 0.75 mM K_2SO_4, 0.1 mM KCl plus 0.75 mM K_2SO_4, and 5.0×10^{-5} mM $(NH_4)_6Mo_7O_{24}$ significantly decreased nitrate concentration in expanded leaves of lettuce, and tap water and distilled water did slightly. However, all treatment solutions did not significantly lower the nitrate concentration in petioles (Table 2).

The second experiment showed that all KCl, K_2SO_4 and $(NH_4)_6Mo_7O_{24}$ contained solutions reduced nitrate concentration both in leaves and petioles, but increased concentration of treatment solutions did not enhance the efficiencies (Table 3). In conclusion, 0.1 mM KCl, 0.75 mM K_2SO_4, 0.1 mM KCl plus 0.75 mM K_2SO_4 and 5.0×10^{-5} mM $(NH_4)_6Mo_7O_{24}$ were efficient treatment solutions in decrease nitrate concentration of hydroponic lettuce for short-term treatment, but their efficiencies were not improved with the increased solute concentration.

In additionally, a glasshouse experiment was conducted to investigate the effects of three kinds of nitrogen-free solutions on Vc and nitrate contents in expanded leaves and old leaves during eight days (Liu et al., 2012b). The results showed that using nitrogen-free solutions could decrease nitrate content of expanded leaf blades, petioles and old leaves (Table 4). The expanded leaf blades are more sensitive to nitrogen disruption treatments in nitrate reduction, followed by petioles and old leaves.

Table 2. Effects of different treatment solutions on nitrate reduction in leaves and petioles of lettuce

Prior to harvest treatment solutions	Nitrate concentration in expanded leaves (mg/kg)	Nitrate concentration in petioles (mg/kg)
Full nutrient solution	1800a	2094a
0.1 mM KCl	1101cd	1755a
0.75 mM K_2SO_4	1271bcd	1991a
0.1 mM KCl+0.75 mM K_2SO_4	1084cd	1845a
5.0×10^{-5} mM $(NH_4)_6Mo_7O_{24}$	1203cd	1767a
Tap water	1411abcd	1714a
Distilled water	1520abc	2085a

Note: In the same column, the different letters following the averages represent significant difference at $P<0.05$.

Table 3. Effects of treatment solution concentrations on nitrate reduction in leaves and petioles of lettuce

Treatment solutions	Nitrate concentration in expanded leaves (mg/kg)	Nitrate concentration in petioles (mg/kg)
Full nutrient solution	2538a	3331a
0.1 mM KCl	582b	1703b
0.2 mM KCl	629b	1310b
0.75 mM K_2SO_4	746b	1311b
1.5 mM K_2SO_4	694b	1658b
0.1 mM KCl+0.75mM K_2SO_4	667b	1916b
5.0×10^{-5} mM $(NH_4)_6Mo_7O_{24}$	548b	1845b
5.0×10^{-4} mM $(NH_4)_6Mo_7O_{24}$	887b	1893b

Table 4. Nitrate contents of expanded leaf blades and petioles of lettuce treated with three hydroponic nitrogen-free solutions at three sampling times (mg/kg)

Treatments	Two days		Six days		Eight days	
	Leaf blade	Petiole	Leaf blade	Petiole	Leaf blade	Petiole
Full nutrient solution	3538.7a	3030.9a	1161.9a	2775.6a	1213.6a	2118.8a
Nitrogen-free nutrient solution	1843.2b	1935.3a	434.9b	2239.1ab	377.7b	731.4b
Distilled water	2674.6ab	3143.6a	670.8b	1948.6b	346.7b	1000.2b
KCl solution	3357.7a	3305.4a	539.0b	1571.6b	230.4b	633.7b

Table 5. Vc content in expanded leaf blades of lettuce treated with three hydroponic nitrogen-free solutions (mg/g)

Treatments	Treatment time			
	Two days	Four days	Six days	Eight days
Full nutrient solution	0.290b	0.150b	0.331ab	0.197b
Nitrogen-free nutrient solution	0.365ab	0.310ab	0.287b	0.357a
Distilled water	0.414a	0.498a	0.417a	0.398a
KCl solution	0.417a	0.386a	0.388ab	0.386a

In addition, three nitrogen-free solution treatments increased the Vc content of expanded leaf blades (Table 5), and occasionally the old leaves. The practical significances of distilled water and KCl solution were higher in term of their efficiencies in improving lettuce quality and the low cost.

The results also showed that there were significant positive correlations between nitrate contents of expanded leaf blades, petioles and old leaves (n=36), but no significant correlation was found between Vc contents of them, as well as between Vc contents and nitrate contents either of expanded leaf blades or old leaves.

1.2. Short-Term Continuous Lighting Method

Short-term continuous lighting method is referred to irradiate hydroponic leafy vegetables at certain light intensity with 24 hours photoperiod during 3-5 days before harvest. Effect of light intensity on quality of hydroponic lettuce (*Lactuca sativa* L.) under pre-harvest short-term continuous light (PSCL) was studied by growing lettuce under 48h continuous illumination delivered by red and blue light-emitting diodes (Zhou et al., 2012a). The treatments are listed in Table 6. Four light intensity treatments, including 50, 100, 150 and 200 μmol·m-2·s-1, were designed. Results showed that nitrate concentration in lettuce shoot decreased significantly after being treated with 48h continuous lighting, while the content of soluble sugar and vitamin C increased substantially (Table 7). It was observed that the effect of PSCL on improving lettuce quality was greatly affected by light intensity. The decrement in nitrate concentration and increment in soluble sugar and Vc content were quite low at light intensity of 50 μmol·m-2·s-1, and gradually increased as light intensity increased from 50 μmol·m-2·s-1 to 200 μmol·m-2·s-1, however, the marginal benefit of light intensity to lowering nitrate concentration and increasing Vc content declined rapidly when light intensity increased beyond 100 μmol·m-2·s-1. In conclusion, the PSCL is an effective method to improve lettuce quality and the light intensity of 100 μmol·m-2·s-1-150 μmol·m-2·s-1 is economically optimal. Furthermore, effect of pre-harvest continuous light with different red/blue ratio on photon flux density (R/B ratio) on reducing nitrate accumulation was studied by growing lettuce (*Lactuca sativa* L.) under continuous illumination delivered by light-emitting diodes (Zhou et al., 2012b). The treatments are listed in Table 8.

Table 6. Light intensity and spectral composition in different treatments

Treatment	Total light intensity (μmol·m^{-2}·s^{-1})	Spectral composition (μmol·m^{-2}·s^{-1})	
		Red	Blue
LED1	50	40	10
LED2	100	80	20
LED3	150	120	30
LED4	200	160	40

Table 7. Initial and terminal fresh weight, nitrate concentration, soluble sugar and Vc content in lettuce

Time	Leaf blade			Petiole		
	Nitrate concentration (mg·kg^{-1})	Soluble sugar content (%)	Vc content (mg g^{-1})	Nitrate concentration (mg·kg^{-1})	Soluble sugar content (%)	Vc content (mg g^{-1})
Initial	2173.5a	0.96b	0.242b	4701.1a	0.62b	0.083b
Terminal	1586.9a	2.34a	0.440a	3434.9b	2.23a	0.123a

The initial value are the means of three plants measured at the beginning of continuous light and the terminal value are the means of twelve plants measured at the end of continuous light. Different letters in the same column show significant difference at *P*<0.05, by LSD.

Table 8. Photon flux densities in each spectral component in different treatments

Treatments	R/B ratio	Photon flux density of red and blue component ($\mu mol \cdot m^{-2} \cdot s^{-1}$)	
		Red	Blue
LED1	2	100	50
LED2	4	120	30
LED3	8	133	17
LED4	--	150	0

Table 9. Initial and terminal fresh weight, nitrate concentration, and soluble sugar content in lettuce. The initial value are the means of three plants measured at the beginning of continuous light and the terminal value are the means of twelve plants measured at the end of continuous light

Time	Leaf blade			Petiole		
	Fresh weight (g)	Nitrate concentration ($mg \cdot kg^{-1}$)	Soluble sugar content (%)	Fresh weight (g)	Nitrate concentration ($mg \cdot kg^{-1}$)	Soluble sugar content (%)
Initial	10.7b	3125.7a	0.14b	7.09a	5099.6a	0.47b
Terminal	15.2a	1317.8b	2.11a	8.36a	3660.7b	2.06a

Different letters in the same column show significant difference at $P<0.05$, by LSD.

Results show that nitrate concentration decreased by 1648.0-2061.1 mg•kg-1 in leaf blade and 962.9-2090.3 mg•kg-1 in petiole, accompanied by a dramatic increase in soluble sugar content.

Compared with monochromatic red light treatment, the decrease in nitrate concentration and increase in soluble sugar content in lettuce under mixed red and blue light were more pronounced (Table 9).

The lowest nitrate concentration was observed in the treatment with R/B ratio of 4. It's concluded that pre-harvest exposure to 48h continuous LED light could effectively reduce nitrate accumulation in lettuce and this process is strongly affected by R/B ratio of light. This study may provide new perspective for pre-harvest quality management of vegetable, especially in commercial leaf vegetable production under artificial lighting.

CONCLUSION

Nitrogen deprivation method and short-term continuous lighting method are two established short-term methods to improve nutritional quality of hydroponic leafy vegetables before harvest. Therefore, the regulation strategy for promoting nutritional quality of hydroponic leafy vegetables before harvest can be used in practice. With this strategy, high productivity and high-quality leafy vegetables can be gained through short-term treatment method before harvest.

REFERENCES

Chung, S. Y., Kim, J. S., Kim, M., Hong, M. K., Lee, J. O., Kim,C. M., Song, I. S. Survey of nitrate and nitrite contents of vegetables grown in Korea. *Food Additives and Contaminants*, 2003, 20:621–628.

Eichholzer, M., Gutzwiller, F. Dietary nitrates, nitrites, and N-Nitroso compounds and cancer risk: a review of the epidemiologic evidence. *Nutrition Review*, 1998,56,95-105.

Liu, W. K., Qichang Yang, Zhiping Qiu. Spatiotemporal changes of nitrate and Vc contents in hydroponic lettuce treated with various nitrogen-free solutions. *Acta Agriculturae Scandinavica, Section B - Soil and Plant Science*, 2012b, 62 (3):286-290.

Liu, W. K., Yang, Q. C. Short-term treatment with hydroponic solutions containing osmotic ions and ammonium molybdate decreased nitrate concentration in lettuce. *Acta Agriculturae Scandinavica, Section B - Soil and Plant Science*, 2011, 61(6): 573-576.

Liu, W. K., Yang, Q. C., Du, L. F. Effects of short-term treatment with various light intensities and hydroponic solutions before harvest on nitrate reduction in leaf and petiole of lettuce. *Acta Agriculturae Scandinavica, Section B - Soil and Plant Science,* 2012a, 62 (2): 109-113.

Maff, U. K. 1999. Nitrate in lettuce and spinach. In: *Food surveillance* information sheet Number 177, edited by Joint Food Safety and Standards Group: http://www.maff.gov.uk/food/infsheet/1999/no177/177nitra.htm.

Santamaria, P. Nitrate in vegetables: toxicity, content, intake and EC regulation. *J. Sci. Food Agric.*, 2006, 86, 10-17.

Sohn, S. M., Yoneyama, T. Yasai no shou-san: Sakumotsu-tai no syou-san no seiri, syuu-seki, hito no sessyu (Nitrate in Vegetables: Nitrate physiology and accumulation in crops, and human intake). *Nogyo oyobi Engei*, 1996, 71, 11, 1179-1182.

Sušin, J., Kmecl, V., Gregorčič, A. A survey of nitrate and nitrite content of fruit and vegetables grown in Slovenia during 1996–2002. *Food Additives and Contaminants* , 2006, 23: 385–390.

Zhong, W. K., Hu, C. M., Wang, M. J. Nitrate and nitrite in vegetables from north China: content and intake. *Food Additives and Contaminants*, 2002,19: 1125-1129.

Zhou, W. L., Liu, W. K., Yang, Q. C. Quality changes of hydroponic lettuce under pre-harvest short-term continuous light with different intensity. *The Journal of Horticultural Science and Biotechnology*. 2012b, In press.

Zhou, W. L., Liu, W. K., Yang, Q. C. Reducing nitrate concentration in lettuce by elongated lighting delivered by red and blue LEDs before harvest. *Journal of Plant Nutrition*, 2012a, In press.

Zhou, Z. Y., Wang, M. J., Wang, J. S. Nitrate and nitrite contamination in vegetables in China. *Food Review International*, 2000,16: 61-76.

INDEX

A

acetaminophen, 120, 124
acetic acid, 67
acetone, 67, 75
acid, 49, 54, 67, 74, 77, 78, 82, 89, 99, 103, 111, 113, 119, 131, 134
adaptability, 1, 2, 5, 6, 7, 8, 9, 15, 23, 25, 26, 27, 40, 50, 86
adaptation, 2, 3, 5, 23, 24, 32, 38, 43, 51, 135
adhesion, 74, 120, 125
Africa, 50, 53, 99, 134
aggregation, 121
aggressiveness, 43, 44
aging process, 119
agriculture, 1, 3, 6, 9, 12, 40, 42, 44, 49, 53, 55
agri-firms, 20, 22, 24, 26, 27, 40
agroecosystems, 43, 44, 45, 50, 53
air temperature, 87
alanine, 131
albumin, 74
alfalfa, 132
algae, 81
alimentation, 117
allele, 101, 106, 107, 108
allergens, 103, 113
alopecia, 103, 114
alternative treatments, 118
amino acid(s), 51, 101, 102, 111, 114, 118, 131, 134, 138
ammonium, 140, 154
anaphylaxis, 123
anatomy, 60
anemia, 123
animal welfare, 5, 10, 18, 35
ANOVA, 90, 145
anticancer activity, 120
anticancer drug, 77

anti-inflammator, 64, 65, 76, 78, 79, 80
anti-inflammatory agents, 78
antioxidant, 119, 120, 121, 124, 127, 148
antioxidative activity, 103, 111
antipyretic, 83
apoptosis, 75, 120, 125, 127
appetite, 113
Argentina, 43, 44, 45, 46, 47, 48, 49, 50, 52, 53, 54, 57, 58, 62
arginine, 101, 111, 118
arteriosclerosis, 121
artery, 120
arthropods, 132
Asia, 46, 53, 138
assessment, 5, 6, 7, 9, 10, 22, 62, 75, 99
assimilation, 49, 55
asymmetry of information, 11
atherosclerosis, 119, 120, 121
atopic dermatitis, 114
autonomy, 6, 7
awareness, 134, 136

B

β-conglycinin, ix, 101, 102, 103, 104, 105, 106, 107, 109, 110, 111, 112, 113, 114, 115
Baars, 86, 88, 95, 96, 97, 100
Bachev, 2, 3, 5, 6, 7, 8, 11, 12, 18, 20, 21, 41, 42
backcross, 103
bacteria, 122, 126
base, 47, 50, 64, 78, 83, 103, 130
Beijing, 147
beneficial effect, 126
beneficiaries, 5, 29, 38
benefits, 4, 6, 19, 20, 21, 29, 81, 96, 103, 111, 118, 124, 129, 134, 135, 136
Benson, 1, 42
benzene, 67

bilateral, 40
biochemistry, 113
biodiversity, 4
biological activities, 138
biomass, 48, 49, 51, 58, 59, 60, 132
biomass growth, 60
biopsy, 75
biotypes, 44, 55, 62
bleaching, 65
blood, 69, 119, 120, 121, 123, 125, 126, 134
blood pressure, 120, 121, 125, 126
blood urea nitrogen, 123
body fat, 103, 110, 112
bone, 134, 137
border control, 3
brain, 78, 138
Brazil, 117
breast cancer, 74, 120, 137
Breece, 42
breeding, 53, 54, 101, 103, 108, 110, 130, 134
bronchial asthma, 123
Bruker DRX, 67
Bulgaria, 1, 2, 5, 6, 10, 11, 12, 13, 14, 15, 16, 22, 23, 24, 25, 26, 27, 29, 31, 33, 36, 37, 39, 41, 42
Bulgarian farms, vii, 1, 2, 7, 8, 17, 28, 30, 31, 32, 33, 35
bureaucracy, 21
business farm, 1, 2, 20, 21

C

cabbage, 149
calcium, 102, 118, 129, 140, 141
calcium carbonate, 140
cancer, 64, 69, 74, 75, 76, 77, 82, 119, 120, 125, 130, 134, 148, 154
cancer cells, 69, 74, 75, 120
cancer progression, 120
CAP, 1, 2, 7, 9, 10, 27, 28, 29, 30, 31, 32, 33, 34, 35, 36, 37, 38, 39, 40
carbohydrate(s), 46, 50, 55, 56, 79, 118, 148
carcinogen, 120, 123
carcinogenesis, 120, 123, 127
cardiovascular disease, 119, 120, 129, 130, 131, 134, 135
cardiovascular disease, 119, 120, 131, 135
cardiovascular morbidity, 121
cardiovascular risk, 120, 126, 127
case study, 42
cattle, 29, 53, 96, 99
cDNA, 105, 106
CEE, 42
cell biology, 113

cell cycle, 75
cell death, 75, 120
cell line, 64, 74, 75, 76
cellulose, 89, 91, 96
ceramide, 63, 68, 78
cerebroside, 63, 67, 68, 78
cerebrovascular disease, 119
cervical cancer, 74
chemical, 49, 54, 55, 56, 59, 67, 75, 85, 86, 88, 89, 90, 98, 99, 100, 119, 121, 123, 138
chemical reactions, 119
chemical structures, 67
chemoprevention, 123
chemotherapeutic agent, 123
chemotherapy, 64, 65, 75
chicken, 133
China, 130, 133, 136, 147, 148, 154
chloroform, 67
chlorophyll, 55, 62
cholesterol, 102, 120, 121, 125, 126, 134
chromatography, 67, 69, 80, 81
chromosome, 84, 102, 104, 105, 106, 111
climate change, 5
clinical trials, 124
clone, 46, 105
CO2, 55, 75
coalition, 3, 18, 20, 21
coenzyme, 125
collagen, 121
colon, 74, 120
colorectal cancer, 120, 127
commercial, 11, 19, 22, 28, 35, 49, 55, 73, 125, 136, 153
communities, 44, 53
community, 3, 44, 45, 53, 58
comparative advantage, 19, 21
compatibility, 86
compensation, 109
competition, 5, 6, 9, 18, 20, 23, 32, 38, 46, 48, 49, 51, 52, 54, 55, 56, 58, 59, 98, 99
competitive advantage, 40
competitiveness, vii, 1, 2, 3, 4, 5, 6, 7, 8, 9, 10, 22, 23, 25, 26, 27, 35, 36, 38, 39, 40, 42
competitors, 3, 4
compilation, vii
composition, 44, 58, 94, 99, 100, 103, 104, 111, 115, 131, 136, 137, 152
compounds, 63, 64, 65, 66, 67, 68, 69, 70, 71, 72, 73, 74, 75, 76, 77, 78, 79, 80, 82, 117, 118, 119, 120, 122, 125, 127, 154
conditioning, 44
conductance, 48, 55
conference, 98, 100

Congress, 41, 42, 99
consolidation, 19, 21
constituents, 81, 118, 122, 124
consulting, 21
consumers, 118, 129, 131, 132, 133, 134, 135, 136
consumption, 2, 18, 20, 44, 120, 123, 124, 126, 130, 133, 134, 138, 148
contact dermatitis, 123
contamination, 61, 154
contradiction, 124
control measures, 3
controversies, 124
cooperation, 4, 6, 18
cooperatives, 2, 6, 13, 18, 19, 20, 22, 24, 26, 40
correlation, 64, 103, 110, 115, 151
corruption, 20, 29
cortex, 51
cosmetics, 65, 79
cost, 3, 52, 86, 151
cotton, 44, 49, 139, 140, 141, 142, 143, 144, 145, 146
cotyledon, 131, 136
creatinine, 123
creep, 45, 46, 50
critical period, 52
crop rotations, 44
crop(s), 2, 10, 12, 14, 15, 16, 17, 18, 20, 22, 23, 24, 25, 26, 28, 29, 30, 31, 32, 33, 34, 35, 36, 40, 41, 43, 44, 45, 47, 49, 52, 53, 56, 57, 61, 134, 136, 148, 154
crude oil, 65
cultivars, 54, 79, 108, 131, 132, 135, 136, 138, 146
cultivation, 19, 147, 148, 149
cultural heritage, 4
cultural practices, 56
culture, 147, 148
curcumin, 65, 80
cysteine, 119, 121, 122
cytochrome, 122, 123, 126
cytotoxicity, 75, 82

D

Davidova, 5, 42
declining markets, 6
defects, 101, 104
deficiencies, 108, 111
deficiency, 99, 101, 103, 104, 106, 107, 108, 110, 111, 112, 114
deformability, 126
degenerative conditions, 119
degradation, 100, 107, 122
degumming, 65, 66, 78

dehiscence, 54
Delgado, 1
dementia, 125
Denmark, 41
deoxyribose, 78, 83
dependency, 6, 14, 19, 26
deprivation, xi, 147, 148, 149, 150, 153
depth, 56
derivatives, 65, 80, 121
detectable, 75
detection, 44, 45, 105, 106
detoxification, 122, 127
developing countries, 1, 86
diabetes, 126
diet, 113, 118, 123, 125
dietary supplementation, 119
diffusion, 52
digestibility, 97, 100
digestion, 99
dimerization, 127
diodes, 152
direct foreign investment, 22
direct payment, 27, 29, 35
disclosure, 139, 145
diseases, 86, 117, 118, 119, 148
dislocation, 76
dispersion, 43
distillation, 121
distilled water, 67, 150, 151
distribution, 50, 51, 52, 55, 59, 86
diversification, 4, 12, 37
diversity, 44, 58, 59, 112, 130
division of labor, 21
DNA, 63, 64, 65, 69, 70, 71, 72, 73, 74, 75, 77, 78, 79, 80, 81, 82, 83, 84, 106, 107, 112, 119, 120
DNA polymerase, 63, 64, 79, 80, 81, 82, 83, 84
dominance, 50, 89, 96
donations, 79
dosage, 81, 124
Drosophila, 69, 81, 82
drought, 48, 50, 59
drugs, 77
dry matter, 54, 86, 88, 97, 100
drying, 88, 89

E

Eastern Europe, 42
Eastridge, 42
ecology, 60, 100
economic crisis, 5
economic efficiency, 17, 31
economic performance, 20

economies of scale, 2, 19, 21, 35
eco-tourism, 2
edamame, 129, 130, 131, 132, 133, 134, 135, 136, 137, 138
edema, 64, 76, 82
EEA, 60
efficiency, 1, 2, 3, 4, 5, 6, 7, 8, 9, 14, 18, 19, 20, 23, 24, 26, 27, 40, 42
Egypt, 117, 139, 140
electron, 47
electrophoresis, 104
elongation, 50, 54
elucidation, 81, 105
emigration, 18
emotion, 134
employees, 20
employment, 4, 6, 11, 12, 19, 20
employment opportunities, 6
encoding, 77, 103, 115
endothelial cells, 74, 119
endothelium, 121
energy, 44, 99, 148, 149
energy consumption, 44
entrepreneurs, 18, 19, 20
environment, 2, 3, 4, 5, 6, 7, 8, 19, 23, 25, 26, 27, 32, 38, 40, 45
environmental aspects, 35
environmental conditions, 44, 50
environmental factors, 135
environmental management, 19
environmental protection, 10
environmental stress, 59
environmental sustainability, 10, 33, 35
environments, 43, 46, 51, 57, 122, 126
enzyme, 64, 65, 69, 70, 71, 72, 74, 78, 80, 81, 82, 83, 84, 102, 113, 127, 133
enzymes, viii, 64, 70, 73, 74, 118, 119, 122, 123, 125
epidemic, 62
epidemiologic, 154
epinephrine, 121
equipment, 11, 13, 56, 57
erosion, 44, 58, 86
erythrocytes, 126
essential fatty acids, 134
Ethiopia, viii, 85, 86, 87, 92, 93, 94, 95, 96, 97, 98, 99, 100
ethnicity, 134
eukaryotic, viii, 63, 64, 65, 66, 67, 71, 72, 74, 77, 78, 79, 80, 82
eukaryotic cell, 79
Europe, 5, 41, 42, 53
European market, 9

European Union (EU), v, vii, 1, 2, 5, 9, 10, 11, 23, 27, 28, 29, 30, 31, 32, 33, 35, 36, 37, 40, 41, 42
evaporation, 67
evidence, 118, 121, 122, 154
evolution, 1, 2, 3, 5, 7, 31, 40, 45, 49, 61, 113
excision, 64, 78, 83
excretion, 123
exercise, 126
exonuclease, 82
expenditures, 4
expertise, 21, 28, 31, 32, 33, 35
export subsidies, 28
exports, 10
exposure, 53, 122, 153
extracellular matrix, 120
extraction, 65
extracts, 52, 63, 117, 119, 120, 122, 127

F

families, 65, 71, 77, 113
family members, 4, 12, 18, 21
farm efficiency, 2, 3, 4, 6, 9, 40
farm size, 18
farmers, 1, 4, 85, 131, 133, 137
farmland, 10, 11, 13, 18, 19, 20, 22
farms, 1, 2, 3, 4, 5, 6, 7, 8, 9, 10, 11, 12, 13, 14, 15, 16, 17, 18, 20, 21, 22, 23, 24, 25, 26, 28, 29, 30, 31, 32, 33, 34, 35, 36, 37, 38, 39, 40, 42, 85
fat, 103, 125
fatty acids, 77, 82, 134
feeding value, 97
feedstuffs, 86
fertilization, 140, 145, 146
fiber, 94, 97, 98
fiber content, 98
fibroblasts, 74
fidelity, 80, 81
field crops, 10, 25, 26, 28, 29, 30, 31, 32, 33, 34, 35, 36, 40
filtration, 65
finance supply, 8
financial, 3, 6, 7, 23, 26, 38, 79
firm, 4, 5, 18, 21, 22
fish, 71
flavor, 118, 130, 133, 134, 136
floods, 23
flora, 44, 57, 61
Fodder Trees, 100
food, 2, 4, 9, 12, 18, 37, 38, 41, 42, 65, 76, 79, 80, 98, 102, 103, 110, 111, 113, 122, 129, 130, 134, 138, 154
food chain, 2, 4

food products, x, 9, 129
food safety, 37, 38
food security, 98
forage crops, 18
foreign investment, 22
formal, 18, 21, 40
formation, 21, 45, 47, 49, 102, 107, 114, 115, 119, 123
fragments, 45, 50, 52, 56, 58
free radicals, 119, 122
frozen produce, ix, 129
fruits, 11, 50
functional food, 79, 137
funding, 8, 21, 23, 25, 26, 27
fungi, 122

G

gastric mucosa, 123
gel, 67, 102, 104, 110, 113, 114
gel formation, 102
gelation, 102, 110, 111, 113, 114, 115
gene expression, 104, 105, 111, 125
gene silencing, ix, 101, 107
genes, 101, 103, 104, 105, 106, 107, 111, 112, 113, 114
genetic background, 108
genetic diversity, 59
genome, 65
genotype, 104, 112, 131
genus, 45, 47, 53, 61, 125
Georgia, 131, 137
germination, 47, 54, 78, 83, 131
global climate change, 5
global warming, 5
globalization, 5
glucose, 68, 75, 77, 78, 79, 131, 134
glucoside, 63, 67, 68, 77, 81, 83
glutamine, 75
glutathione, 119, 123, 124
glycerol, 69
glycinin, ix, 101, 102, 103, 108, 109, 110, 111, 112, 113, 114, 115
glycol, 80
glycoside, viii, 63, 67, 68, 77, 78, 135
glyphosate, vii, viii, 43, 44, 45, 49, 52, 55, 58, 62
Gortona, 5, 42
governance, 2, 3, 6, 7, 8, 10, 19, 24, 31, 32
governance costs, 3
Gracia, 42
grain size, 54
grass, 46, 50, 56, 57, 59, 60, 85, 86, 87, 88, 89, 90, 91, 93, 95, 96, 97, 98, 99, 100

grasslands, 53
grass-legume, viii, 85, 86, 87, 88, 89, 90, 91, 95, 97, 99, 100
grazing, 10, 23, 24, 28, 29, 30, 34, 36, 40, 85, 97, 98
green alga, 81
greenhouse, 149
growth, 20, 29, 40, 43, 45, 46, 47, 48, 49, 50, 51, 52, 53, 54, 57, 58, 59, 60, 75, 82, 95, 96, 97, 110, 119, 120, 122, 130, 134, 140, 142, 146, 148
growth factor, 44, 50, 51, 96
growth rate, 48, 49

H

H. pylori, 122
habitats, 45
hardness, 102, 109, 115
harvesting, 56, 88
hazards, 55
health, ix, 79, 103, 111, 118, 119, 125, 129, 134, 135, 136, 137, 138, 148
health effects, 125, 137
health promotion, 79
heart disease, 134
hemicellulose, 91, 96
hepatocarcinogenesis, 123, 126
hepatocytes, 120, 121, 122, 127
hepatoma, 121
herbicides, 44, 47, 49, 52, 56, 57, 60, 132
heredity, 108
hermaphrodite, 46
hexane, 65, 67
high blood pressure, 121
high density lipoprotein, 124
highlands, 97, 98, 100
homeostasis, 48
hormones, 134
horticultural crops, 147
human body, 148
human capital, 18
human genome, 65
husbandry, 98
hybrid, 40, 145
hybridization, 104
hydroponics, 147, 148
hydroxyl, 64, 77
hygiene, 10
hypertension, 121, 124
hypocotyl, 79
hypotensive, 113

I

identification, 3, 47, 79, 83
illumination, 152
immune function, 138
immune system, 134
immunity, 119
imported products, 9, 38, 136
in vitro, 63, 76, 78, 79, 82, 97, 100, 122, 126
in vivo, 64, 76, 119, 122
income, 1, 2, 4, 5, 10, 12, 18, 20, 28, 29, 30, 31, 36, 38, 40, 134
income support, 4
independence, 6, 23, 38
individuals, 3, 18, 20, 45, 46, 121, 123, 126
induction, 119, 120, 125
industries, 3, 18, 137
industry, 130, 136
infants, 148
inflammation, viii, 64, 65, 76, 78, 79
inflammatory responses, 80, 84
inflation, 18
informal, 40
informal sector, 35
ingestion, 123, 126
inheritance, 108, 112
inhibition, 47, 54, 55, 64, 65, 71, 72, 74, 75, 76, 77, 78, 80, 113, 120, 122, 124, 125
inhibitor, 59, 64, 65, 71, 75, 76, 77, 80, 82, 84, 133
inputs supply, 21
insecticide, 132
insects, 95, 132
insertion, 103
institutional environment, 7, 27
institutional modernization, 40
institutions, 15, 21
insulin, 83
integration, 1, 2, 4, 6, 19, 37, 40, 114
interaction effects, 139, 143, 144
interdependence, 18
interest rates, 38
interference, 48, 51, 52, 54, 62
Internal structure, 112
international competition, 9
international trade, 3, 4
intervention, 27, 40, 75
intron, 106
investments, 4, 20, 22, 35, 36
investors, 137
ions, 67, 150, 154
iron, 118
irradiation, 51, 101, 103, 113, 114, 149, 150
irrigation, 18, 132, 140

ischemia, 124
isoflavone, 131, 135, 136, 137
isolation, 81
isomers, 131

J

Japan, 63, 69, 79, 101, 112, 113, 114, 115, 130, 133, 134, 136, 148
joint ventures, 20, 21

K

Kagatsume, 41
Kenya, 100
kill, 50, 56
Knezevic, 42
Korea, 133, 154
Koteva, 6, 42

L

land supply, 19, 27
Latin America, 59, 60
LDL, 119, 120, 123
leafy vegetables, vii, xi, 147, 148, 149, 150, 152, 153
LED, 153
legume, vii, viii, 85, 86, 87, 88, 89, 90, 91, 93, 95, 96, 97, 98, 99, 100
legumes, 97, 100
Lerman, 5, 42
leukemia, 74
liberalization, 3, 5, 9
life cycle, 5, 7, 46
ligand, 80
light-emitting diodes, 152
lignin, 89, 94
lima bean, 130
linoleic acid, 77, 78, 103
lipid peroxidation, 119, 126
lipid peroxides, 119
lipids, 81, 118, 119, 126
liquid chromatography, 81
liquidate, 18
liquidity, 6, 7
liver, 74, 119, 120, 123
liver cancer, 74, 120
livestock, 2, 5, 6, 7, 10, 11, 14, 15, 16, 17, 22, 23, 24, 25, 28, 29, 30, 31, 32, 34, 35, 36, 37, 38, 39, 40, 41, 86, 98, 100
localization, 59
locus, 104, 105, 108

low temperatures, 50
LSD, 142, 143, 144, 152, 153
luminosity, 133
lung cancer, 74
lymphocytes, 134
lysine, 134

mutant, 82, 101, 103, 104, 105, 108, 109, 111, 112, 114
mutant gene, ix, 101, 104
mutation, 104, 112, 113
mutations, 77, 114
myocardial infarction, 120
myocardial necrosis, 124

M

machinery, 18, 19, 20
magnesium, 102, 118
Mahmood, 1, 42
malnutrition, 86
management, 2, 5, 6, 8, 9, 18, 19, 20, 21, 28, 36, 38, 43, 44, 45, 51, 52, 54, 55, 57, 58, 103, 129, 135, 136, 141, 153
manganese, 118
Manolov, 42
manufacturing, 65
market, 1, 2, 3, 4, 5, 6, 7, 8, 9, 11, 15, 18, 19, 20, 21, 23, 25, 26, 27, 40
marketing, 2, 4, 5, 8, 16, 18, 19, 21, 23, 24, 25, 26, 27, 130, 135
materials, 13, 65
Mathijs, 5, 42
medicine, 117, 118
Mediterranean, 46
melting temperature, 64, 74
membership, 6, 13, 18, 19, 20, 38
membranes, 78
metabolism, 78, 120, 122, 123, 124
metabolites, 47, 135
metabolized, 123
metabolizing, 123, 125
methanol, 67
methodology, 40
mice, 76, 78, 83, 113
microorganisms, 80
migration, 120, 125
Ministry of Education, 79
mode, 3, 12, 18, 19, 21
modernization, 5, 8, 18, 21, 28, 40
moisture, 47, 51, 57, 65
molecular structure, 77, 80
molecular weight, 102, 123
molecules, 64, 74, 112, 115, 119
morphology, 51, 120
mortality, 120, 121
mRNA, 101, 104, 106
mtDNA, 81
mucosa, 125
mutagenesis, 77, 82

N

NaCl, 133
Nanseki, 5, 42
National Research Council, 43
natural resources, 8, 10, 16, 24
nature, 15, 25, 26
negative effects, 38, 41
Netherlands, 99
neural development, 78
neurogenesis, 83
neutral, 28, 30, 32, 33, 34, 35, 36, 40, 46, 94
New Zealand, 99
nitrates, 154
nitrite, 148, 154
nitrogen, 58, 88, 89, 96, 97, 98, 111, 139, 142, 145, 146, 147, 148, 149, 150, 151, 154
nitrosamines, 123
nodes, 50, 53
nodules, 96, 97
non-smokers, 119
North Africa, 53
North America, 59, 130
Northern blot, 78, 107
nucleic acid, 74
nucleotides, 107
nucleus, 75
nutrients, 46, 51, 79, 80, 137, 140, 147, 151
nutrition, 83, 86, 100, 130, 133, 135, 138, 149

O

oil, 18, 56, 57, 63, 65, 66, 67, 71, 77, 78, 79, 80, 83, 99, 117, 118, 121, 122, 125, 126, 129, 130, 131, 134
oil production, 65, 66
opportunities, 6, 11, 19, 21, 30
optimization, 5
organic matter, 44, 89, 140
organism, 45
organization, 3, 4, 5, 6, 9, 11, 18, 19, 20, 21
organize, 3, 19
organs, 45, 46, 47, 50, 52, 54
osteoporosis, x, 129, 130, 131, 135

ovarian cancer, 135
ownership, 6, 18, 21
oxidation, 113, 119, 123, 126
oxidative stress, 84, 119, 120, 123, 124, 125, 126

P

pachytene, 83
Pakistan, 42, 60, 61
parenchyma, 51
parkinsonism, 81
partial thromboplastin time, 121
pasture, 56, 85, 86, 88, 90, 95, 96, 97, 98
pathogenesis, 120
pathogens, 122
pathways, 112
PCR, 105, 106, 107, 112
Peeters, 42
penicillin, 74
peptide, 102, 113, 114, 134, 138
peptides, 102, 112, 113, 114
percolation, 65
perennial weed, 43, 45, 46, 52, 57
peripheral blood, 69
peroxidation, 103
personal characteristics, 6
personal relations, 21
pests, 86, 137
Petiole, 151, 152, 153
pH, 67, 69, 87, 109, 133, 140
phage, ix, 101, 106
pharmaceutical, 81, 135
phenolic compounds, 120, 135
phenotype, 101, 105, 108, 111, 112
phosphate, 78, 83, 88, 141
phosphorus, 87, 118, 135, 141
photosynthesis, 55
physical properties, 114
physical structure, 44
Physiological, 61, 115
physiology, 43, 45, 154
phytosterols, 83, 118
pigs, 10, 23, 24, 28, 29, 32, 34, 36, 40
placebo, 127
plant growth, 96, 97, 140, 145
plants, 5, 45, 46, 47, 48, 50, 54, 55, 56, 59, 78, 98,
    104, 112, 117, 118, 132, 139, 140, 141, 143, 152,
    153
plaque, 119
plasma levels, 121
plasticity, 44, 48, 51, 53, 57
platelet aggregation, 120, 124
pleiotropy, 104

policy, 19, 28, 29, 30, 31, 32, 34, 36, 38, 148
political power, 21
pollination, 47
polyacrylamide, 104
polymerase, viii, 64, 70, 73, 74, 79, 80, 81, 82, 83,
    84
polymorphisms, 105
polynucleotide kinase, 64, 70, 73, 74, 82
polypeptide, 102
polyphenols, 118
polysaccharides, 100
polyunsaturated fat, 134
polyunsaturated fatty acids, 134
Popovic, 1, 42
population, 18, 43, 44, 45, 49, 52, 53, 55, 106
positive correlation, 64, 110, 151
potassium, 118, 126, 139, 140, 141, 145, 146
Pouliquen, 1, 42
poultry, 10, 14, 15, 16, 17, 20, 22, 23, 24, 28, 29, 32,
    34, 36, 40
preparation, 69, 117, 118, 119, 141
preservation, 4, 5, 18, 37
prevention, 117, 129
privatization, 6, 10, 21, 22
process control, 149
producers, 26, 28, 86, 118, 130, 133
production costs, 3
productivity, 2, 3, 5, 6, 7, 9, 21, 23, 25, 26, 27, 42
profit, 4, 6, 18, 19, 20, 21, 36, 38
profitability, 2, 3, 4, 5, 6, 7, 19, 23, 25, 26, 27, 31,
    38, 135
proliferation, 64, 74, 75, 77, 78, 82
proline, 51
promoter, 78, 82, 84, 104
property rights, 3, 20, 40
prostate cancer, 129
protected areas, 29, 30
protection, 3, 10, 18, 19, 119, 124
protein components, 114
proteins, 51, 65, 101, 102, 105, 111, 112, 113, 114,
    115, 118, 119
prothrombin, 121
public health, 148
public intervention, 40
public support, 3, 5, 19, 21, 28, 29, 30, 32, 36, 38
pulmonary arteries, 121
purification, 66, 71, 78, 82
pyrimidine, 77

Q

quality improvement, 137
Queensland, 99

R

Rader, 42
radiation, 44, 77, 119
radiation therapy, 77
rainfall, 87, 132
Ramadan, 83
rape, 11
reactive oxygen, 119, 122, 125
reactivity, 126
recessive allele, ix, 101, 108
recommendations, 56, 57, 88, 132
reform, 11
regrowth, 95
regulations, 5, 21
relevance, 124
rent, 19, 20
repair, 63, 64, 71, 78, 82, 83, 84
replication, 64, 78, 79, 82
reproduction, 45
reputation, 3, 4, 19, 21, 22
requirements, 3, 7, 18, 21, 29, 33, 86
researchers, 44, 122, 124
reserves, 50, 56
residuals, 18
residues, 54, 57, 62, 69, 83, 86, 132
resistance, 44, 47, 49, 55, 62, 132
resolution, 67
resources, 2, 3, 4, 5, 8, 10, 11, 12, 16, 18, 19, 20, 21, 24, 33, 40, 46, 84, 86, 96, 97, 100, 149
response, 43, 44, 48, 50, 51, 55, 58, 59, 75, 126
restrictions, 3, 7, 10, 21, 22
restructuring, 5
retirement, 6, 18
rhizome, 46, 47, 50, 51, 52, 56, 58
risk(s), 2, 4, 11, 19, 21, 118, 119, 121, 125, 134, 135, 148, 154
RNA, viii, ix, 64, 70, 73, 74, 82, 101, 106, 107, 108, 112
room temperature, 76
root(s), 46, 47, 48, 49, 51, 52, 54, 55, 59, 60, 96, 97
root growth, 49
root system, 48
rules, 40
rural, 18, 19
rural areas, 18, 19
rural development, 19
rural population, 18
Russia, 149

S

safety, 5, 10, 18, 23, 32, 35, 37, 38
Saha, 42
salinity, 60, 61
saliva, 122
salmon, 69, 70, 71, 82
saponin, 138
sclerenchyma, 51
SDS-PAGE, 104, 105, 108
security, 6, 11, 19, 98
sedimentation, 109
seed, 45, 47, 49, 50, 51, 53, 56, 59, 61, 79, 83, 85, 86, 88, 89, 90, 91, 92, 93, 94, 95, 96, 97, 98, 103, 104, 105, 108, 110, 111, 112, 113, 114, 115, 129, 130, 131, 132, 133, 134, 135, 136, 137, 139, 141, 142, 143, 146
seed cotton, x, 139, 141, 142, 143, 146
seeding, 85
seedlings, 47, 53, 59, 140
segregation, 106, 108
selenium, 118, 120
self-employment, 4
sensitivity, vii, 43, 49, 62
serum, 74, 102, 110, 111, 112, 113, 121, 123, 127
services, 2, 5, 6, 8, 16, 18, 19, 23, 24, 32, 37
shareholders, 20
sheep, 29, 99
shelf life, 133
Shoemaker, 1, 42
shortage, 86, 98, 139, 145
side effects, ix, 117, 124
signals, 107
silica, 67
skin, 84, 120, 123
smoking, 119
SNP, 106
sodium, 49, 104, 118
sodium dodecyl sulfate (SDS), 104
soil erosion, 44, 86
solubility, 109, 115
solution, 75, 111, 140, 147, 148, 150, 151
Sorghum halepense, 43, 45, 46, 47, 48, 49, 58, 59, 60
South Africa, 99
South America, 46, 50, 60
South Asia, 134
sowing, 56, 87, 132, 145
soy sauce, 130
soybean seeds, 101, 112, 114, 130, 134
soybeans, 44, 49, 60, 65, 79, 105, 114, 115, 130, 134, 136, 137, 138
soymilk, 102, 112, 115, 129, 130

specialization, 4, 6, 10, 12, 18, 19, 21, 22, 25, 26, 36, 37
species, 43, 44, 45, 47, 48, 49, 50, 51, 52, 53, 54, 57, 58, 61, 64, 65, 66, 71, 72, 74, 77, 79, 80, 81, 85, 86, 88, 91, 95, 96, 97, 98, 99, 119, 125, 127, 132, 145
spectral component, 153
spermatocyte, 83
sprouting, 47, 60
starch polysaccharides, 100
statistics, 30, 145
sterile, 50
sterols, 78, 81
stigma, 81
stimulus, 5
stock, 75
stomach, 74, 120
storage, 18, 51, 56, 101, 102, 111, 112, 113, 114, 115, 133, 138
strategy use, 54
streptococci, 122, 125
stress, 48, 51, 59, 60, 61, 84, 110
stress factors, 51
structural defects, ix, 101, 104
structure, 3, 5, 6, 19, 27, 28, 30, 40, 44, 65, 68, 71, 77, 80, 81, 83, 107, 108, 113, 138
subsistence, 2, 11, 32, 35
subsistence farming, 32
substrate, 69, 71, 72, 73, 74
sucrose, 111, 130, 133, 135
sulfate, 102, 104, 140, 141
sulfonylurea, 49
sulfur, 102, 118, 120, 125
supplementation, 99, 119, 123, 126
suppliers, 13, 37
supply chain, 42
suppression, 119, 123
surveillance, 154
survival, 45, 49, 54, 82
survival rate, 54
susceptibility, 122
sustainability, vii, 1, 2, 5, 6, 7, 8, 9, 10, 16, 24, 33, 34, 35, 38, 40
sustainable development, 148
Sweden, 99
sweet pea, 130
Swinnen, 5, 42
symptoms, 134
synchronize, 148
syndrome, 104
synthesis, 42, 64, 65, 77, 81, 83, 121, 126
systolic blood pressure, 124

T

Taiwan, 137
target, 78, 107
taxation, 19
technical efficiency, 2, 3
techniques, 37, 38, 46, 52
technology, 3, 30, 148
telomere, 82
temperature, 47, 50, 57, 60, 64, 74, 76
texture, 130, 133, 134, 136, 140
Thailand, 132
therapeutic agents, 117
therapy, 77, 121, 135
thermal treatment, 122
threshold level, 96
thrombin, 121
thrombosis, 120
thymine, 80
thymus, 69, 78, 81, 83
tobacco, 6
tocopherols, 83
tofu, 102, 109, 110, 113, 114, 129
topoisomerases, 64, 73, 74
Tosin, 42
total cholesterol, 123
tourism, 2, 22
toxic effect, 122, 148
toxicity, 121, 154
toxin, 81
TPA, viii, 64, 65, 75, 76, 78
trade, 3, 4, 9, 11, 19, 22
trade liberalization, 9
traits, 44, 46, 50, 52, 57, 58, 130, 131, 137
transaction, 2, 3, 4, 5, 6, 19, 21
transaction costs, 2, 3, 4, 5, 6, 19, 21, 32
transactions, 4, 5, 6, 13, 19, 22
transcription, 105, 106, 107, 120
transcripts, 78, 105
transformation, 8, 21, 44, 111, 118
translocation, 55
transpiration, 48
treatment, 75, 82, 89, 117, 118, 120, 121, 122, 123, 140, 141, 144, 149, 150, 151, 153, 154
trial, 125, 126
triglycerides, 123
trypsin, 133
tryptophan, 134
Tsuji, 41
tumor, 78, 82, 84, 120, 123, 125
Turkey, 145

U

UK, 98, 136, 149
unregistered farm, 11, 13, 18, 23, 25, 26
Uruguay, 46, 50
USA, 69, 74, 82, 83, 84, 98, 99, 100, 145, 146
UV, 77, 119

V

varieties, 5, 21, 50, 108, 111, 114, 129, 130, 131,
    134, 135, 136, 138
vegetables, 10, 11, 12, 22, 23, 24, 25, 26, 28, 29, 30,
    31, 32, 34, 35, 36, 39, 41, 147, 148, 149, 150,
    152, 153, 154
vegetative stages, 54
vein, 74, 77, 121
viscosity, 102
vitamin C, 147, 148, 149, 152
vitamins, 118, 129, 135

W

waste, 44, 63, 65, 66, 67, 71, 77, 78

water, 5, 44, 46, 48, 49, 51, 59, 60, 65, 67, 76, 118,
    119, 120, 130, 150, 151
water absorption, 48, 49
water shortages, 5
weight gain, 99
weight loss, 123
welfare, 5, 10, 18, 35
wholesale, 2
wild type, 104, 108, 109
World Bank, 42

Y

yield, 48, 52, 54, 85, 86, 88, 89, 90, 92, 93, 95, 96,
    97, 98, 99, 100, 103, 108, 110, 131, 132, 135,
    136, 137, 138, 139, 140, 141, 142, 143, 144, 145,
    146, 148, 149

Z

Zawalinska, 1, 5, 42
zinc, 118